Economic Models and Applications
of Solid Waste Management

Economic Models and Applications of Solid Waste Management

by

Hans-Werner Gottinger

University of Maastricht
The Netherlands

and

Oxford Institute for Energy Studies
United Kingdom

Gordon and Breach Science Publishers
New York • Philadelphia • London • Paris • Montreux • Tokyo • Melbourne

Gordon and Breach Science Publishers

Post Office Box 786
Cooper Station
New York, New York 10276
United States of America

5301 Tacony Street, Drawer 330
Philadelphia, Pennsylvania 19137
United States of America

Post Office Box 197
London WC2E 9PX
United Kingdom

58, rue Lhomond
75005 Paris
France

Post Office Box 161
1820 Montreux 2
Switzerland

3-14-9, Okubo
Shinjuku-ku, Tokyo 169
Japan

Private Bag 8
Camberwell, Victoria 3124
Australia

Library of Congress Cataloging-in-Publication Data

Gottinger, Hans-Werner.
    Economic models and applications of solid waste management / by
Hans-Werner Gottinger.
        p.   cm.
    Includes bibliographical references.
    ISBN 2-88124-398-3
    1. Refuse and refuse disposal—Cost effectiveness.   2. Operations
research.   3. Refuse and refuse disposal—Case studies.   I. Title.
TD793.7.G68   1991
363.72'8—dc20   89-23726
                                                                                      CIP

TP

CONTENTS

v

PREFACE

The management of solid and hazardous waste is one of the pressing environmental and social problems of our time. In order to dispose of, treat and recycle the ever-accumulating amount of waste, in the long run, significant changes in consumption and production patterns of an economy are required. However, because of pressing needs, we cannot wait for the long run. In the meantime we need to provide efficient and effective management systems for coping with the waste problem on a regional and local level.

In this endeavor, the siting of waste management facilities (treatment and disposal) as well as the consideration of technological options for treatment and disposal play a crucial role for the following reasons:

1. By creating cost effective waste management operations in a network of waste disposal and treatment facilities, transportation costs would be minimized.

2. Disposal quantities could be reduced by the economic use of advanced technologies for waste treatment (incineration, pyrolysis, composting, etc.).

3. In view of the increasing shortage of waste disposal sites (landfills) and heightened public opposition to newly proposed sites, a shift toward handling waste management as a strategy for resource recovery and conservation is needed.

On the basis of strategic planning for waste management in the regional context there is a broad range of options that originates in design concepts of waste management planning. These include the selection of waste disposal

vii

sites (closing of dumps, opening of new sanitary landfills, expansion of existing landfills); the selection of waste treatment technologies (balancing existing procedures with new procedures compatible with local requests); siting of facilities in consideration of social, environmental and political factors; and determining capacity levels of waste treatment facilities.

One reason for presenting this book is to portray a state-of-the-art approach to economic and operations research modelling in the management of solid and hazardous wastes, and to demonstrate results of this modelling in particular urban or regional waste management situations. The second purpose of this monograph is to provide an international perspective of the waste management problem. This is evident through case studies of major metropolitan areas such as Munich and Nuremberg, FRG, and Boston, Massachusetts, USA.

This book evolved from a series of lectures sponsored by UNESCO and delivered at the International Centre for Theoretical Physics, Trieste, Italy, in 1982; and at Tohoku University, Sendai, Japan, in 1984. Many thanks to the participants of these meetings, too numerous to mention, for their lively and persistent comments that have led to an improvement of the manuscript. I also appreciate comments by Professors Schenkel (Bundesumweltamt Berlin), G. Anandalingam (University of Pennsylvania), D. Brown (University of Virginia), W. Walker (Rand Corporation), M. Sakaguchi (Osaka University), M. Chatterji (SUNY, Binghamton), and J. Pelkmans (EIPA Maastricht). Dr. Kinkeldei from the Office of Environmental Protection (Bay. Landesamt f. Umweltschutz) was very helpful in providing me with pertinent data leading to the case studies.

Finally, I am very much indebted to Mrs. Steinhausen for her excellent editing of the manuscript by going through numerous cycles of revisions and modifications with great patience; and to Ms. Hecht for drawing figures of professional quality. I very much appreciate their help and support through the Fraunhofer Institute for Technological Forecasting (INT) during my tenure there.

FOREWORD

Continued growth in the standard of living around the world has created negative externalities. One of these externalities is the problem of waste material and how it should be dealt with. In the developed countries, this issue has reached a crisis situation; it is affecting the ecosystems of the individual countries concerned, and of the global community as well. The waste problem is just as serious in the big cities of the developing countries, and is intricately interwoven with the problems of hygiene and public health. Since poorer countries are closely following the developmental path of the wealthier countries, the entire world is at risk because of waste generation. Hazardous materials, including high and low level nuclear waste, also pose a serious threat to human health. Since most waste is generated in the major population centers, disposal facilities need to be situated near these locations for technical and economic reasons. The problem of waste disposal is also interlinked with air pollution, water pollution and transportation problems. Therefore, it is necessary to develop a global as well as a regional perspective including technological, social, economic and political factors.

The crucial consideration is, of course, the location of facilities, i.e., landfill sites, incinerators, etc. New ideas need to be developed in location and space economics for facilities we all need but don't necessarily want located near our homes. The problem exists not only in waste disposal, but also in social welfare facilities such as prisons, mental health facilities, halfway houses, psychiatric centers, adult homes, drug rehabilitation centers, and developmental centers.

After a set of possible location centers is identified, allocation, transportation and disposal strategies will need to be developed considering costs, health hazards, transportation, available technology, political realities, etc.

ix

This objective and preferably computerized (for ease of simulation) process will need to be integrated with the general economic system. The very important subject of recycling and environmental education, of course, should also be a part of the analysis.

Professor Hans W. Gottinger has successfully combined all of these factors in his book. Unlike many industrial engineers, he shows a broader perspective beyond optimization and computer algorithm. He goes a step further and shows how his models can be used in real life situations by applying them to two major cities in the world.

I found his book very stimulating and strongly recommend it to all individuals interested in waste management, environmental studies, facilities location, sanitary engineering, regional planning and related areas.

Manas Chatterji
State University of New York at Binghamton

CHAPTER 1

Scope of the Problem

1.1 INTRODUCTION

Waste management today is made difficult and costly by the increasing volumes of waste produced, by the need to control potential serious environmental and health effects of disposal, and by the increasing lack of land in metropolitan areas, partly due to public opposition to proposed sites as locally unwanted land use (LULU). Waste management, once strictly a local and private sector matter, now involves regional, state and federal authorities in a highly complex regulatory framework. Overall growth in (municipal) solid waste generation seems to have slowed in the last years, due to the increased environmental awareness of the consumer and technology driven changes in industry. Yet, an alleviation of the problem of waste generation appears not to be in sight; furthermore, increasing concern is voiced in the national press that the United States is quickly running out of adequate landfill capacity in major metropolitan areas. In a recent article in the Wall Street Journal (April 15, 1986), it is estimated that by 1990 more than half of the U.S. cities will exhaust the existing landfills while 90% of municipal waste is currently dumped in landfills. The consequence so far has been a dramatic increase, sometimes quintupling of disposal costs, within the last five years. Thus, the U.S. faces a major waste management problem, in magnitude and in kind, that has to be solved at acceptable social costs and risks in the years ahead. The problems and constraints applying to non-hazardous solid wastes (garbage) in reducing costs of waste management, also apply, in principle, to "hazardous solid wastes" with the provision that in addition to costs also risks, that is risk related costs pertaining to externalities (e.g. health effects and environmental impacts) should be minimized.

Various legislative initiatives and procedures have been activated within the past few years in the leading industrial countries (OECD) (1981). Here are a few figures which illustrate the extent and importance of the solid waste management problem. In the United States, municipal solid waste (mainly trash and garbage from residential, commercial and industrial sources)

1

amounted to about 140 million metric tons in 1978. (This is enough to fill the Munich Olympic Stadium three times a day throughout the year!) Per capita generation in 1978 was about 650 kg per year, as compared to 460 kg in 1960. The situation is similar in highly industrialized West European countries, but for illustrative purposes we cite the U.S. situation for presenting the major dimensions.

In the United States, roughly 30–40% of industrial wastes generated in 1977 was in solid form, the rest liquids and sludges. Industrial waste generation is growing about 3–4% per year. An increasing percentage of the waste results from pollution control processes. About 10–15% of industrial wastes may be considered potentially hazardous, requiring special safeguards in handling and disposal to reduce the risk of human and environmental health effects. Furthermore, 18,000 municipal waste water treatment plants produce 5 million metric tons (dry weight) of sludge per year. The amount is expected to increase greatly in the coming years due to higher levels of treatment. Agricultural wastes total about 430 million metric tons (dry weight) per year.

Disposal of waste on land can affect public health and environmental quality in many ways. Improper disposal practices have led to death and injuries of workers, direct exposure of nearby residents to toxic wastes, contamination of groundwater and surface streams, air pollution, damage to wetlands and other environmentally sensitive areas, contamination of croplands with heavy metals, and other problems.

Besides land disposal methods incinerators have been increasingly employed to dispose of municipal solid wastes, industrial wastes and sewage sludge. Despite reducing the large bulk of wastes, ash and other residues still have to be landfilled. To some extent, however, the presence of municipal incinerators may conflict with air pollution standards. In the past years a number of other viable waste treatment technologies, in the category of energy recovery and recycling, have proliferated. We shall look at various technology options of waste treatment and disposal in the regional context, considering existing facilities as well as viable, state-of-the-art treatment technologies. Thus we are interested in exploring the minimal cost/risk design of waste management.

Potentially, the most widely significant environmental effect is contamination of groundwater. Inadequate management of hazardous wastes came to public attention in the case of 'Love Canal' at Niagara Falls in the United States. The former Canal site had long been used for burial of industrial chemical wastes prior to 1953. Over 200 families were relocated, extensive removal construction was planned and additional monitoring and testing were taking place. Quite recently, there were exclusive and consistent reports on contamination sites in the U.S. at Times Beach and other places of dioxin poisoning to such an extent that the federal government was forced to purchase the contaminated land.

1.2 DEFINITION OF SOLID WASTE

How are the concepts of 'solid waste' and 'hazardous waste' legally related? According to the U.S. Resource Conservation and Recovery Act (RCRA) of Oct. 1976 hazardous waste is part of the definition of solid waste. 'Solid waste' is defined as 'any garbage, refuse, sludge from a waste treatment plant, or air pollution control facility, and other discarded material resulting from industrial, commercial, mining and agricultural operations, and community activities'.

The phrase in this definition that has attracted the most controversy is 'and other discarded material,' as the door is apparently open to an exceedingly expansive interpretation of what may be included in the definition of solid waste. The U.S. Congress has made confusion complete by defining 'solid' waste as a solid, liquid or gaseous material. This apparent inconsistency provides a prime example of statutory definitions.

Once the U.S. Congress defined solid waste it proceeded with its definition of 'hazardous waste' as follows:

> . . . a solid waste, or combination of solid wastes, which because of its quantity, concentration of physical, chemical or infectious characteristics may (a) cause, or significantly contribute to an increase in mortality or to an increase in serious irreversible, or incapacitating reversible illness; or (b) pose a substantial present or potential hazard to human health or the environment when improperly treated, stored, transported, or disposed of, or otherwise managed.

We will accept this rather broad definition as a guideline at least for the conceptual part of our model building.

1.3 COSTS OF SOLID WASTE MANAGEMENT

Solid waste management systems have three basic cost centers associated with collection, transfer station usage or intermediate waste processing, and ultimate disposal. Intermediate waste processing (IWP) now represents a wide array of available and partially proven technologies ranging from incineration, to conversion (pyrolysis), to composting, to resource recovery (reclamation). Given these sets of different waste management alternatives we can choose among various waste management designs.

Collection costs include the cost of pick-up from household and commercial establishments and of hauling to a transfer station, processing plant or disposal site. Transfer station costs include waste handling and hauling to landfills. Disposal costs are largely landfill costs. For certain communities, however, transfer and disposal costs may exceed cost of collection. The relatively high cost of collection service is largely due to its labor intensiveness, although equipment and fuel costs are also important factors.

The costs of environmentally adequate disposal are substantially in excess of open dumping or inadequate practices. The costs of landfilling have escalated over the past few years in major metropolitan areas, as has the cost of incineration and that of other IWP technologies. Also the relationships of these costs among competing technologies, procedures and practices have changed.

Similarly, as we observe in Western Europe, the thrust of the Japanese solid waste management is towards resource and energy recovery from waste. This policy is necessitated by the shortage of landfill sites in Japan and the high cost of disposal. In particular, the activities have shifted toward municipal resource recovery, primarily because of the cities' need for an alternative land disposal.

In the U.S. approximately 10% of municipal waste was recovered in 1978, 7% for recycling and 3% for energy production. This ratio has not changed significantly since then. This level is low compared with levels achieved by West European countries which range between 20 and 60%. A number of interrelated factors have hampered a more rapid expansion of resource recovery in the U.S. The traditional forms of waste disposal, dumping or landfilling, have generally been cheap, at least in terms of direct costs; environmental damage has long been ignored, and compared with many European countries land has been plentiful. But the situation in major metropolitan areas has changed dramatically over the past decade.

Obviously, the problem could be simplified if the quantity or at least the volume of waste products were reduced by salvaging or waste reclamation. Salvage operations, which were profitable a number of years ago, have largely been abandoned for economic reasons, because of land interference with normal disposal operations. Yet the fact remains that a reduction in the amount of waste that must be handled by municipal or private facilities would result in substantial savings. Because disposal costs are often financed from general funds or lumped with other operating costs, many large producers of waste are not aware of the total costs of refuse disposal. Frequently, as is the case with liquid wastes, a large share of the total disposal costs are born by the community. The result is that there is little or no incentive to reduce the quantity of wastes or find new uses for waste materials.

1.4 NEED FOR REGIONAL MANAGEMENT

The primary reasons to move the responsibility for solid waste management to a regional level instead of a level of local towns and cities, which is the current practice, is economics, but technical and political feasibility is also a consideration.

One could define 'economies of scale' as declining average cost as scale increases, and then establish empirically the existence of economies of scale for transfer station, incineration and landfill operations.

When the management is on a local basis, wastes are collected by 'packers' and hauled directly to the disposal sites. There is a definite scope for improvement here. Packers are specialized units for collection and usually have one third the capacity of trailer trucks. This means the packers go back and forth between waste generation areas and the disposal sites. In addition to incurring heavy transportation costs, this operation requires more packers and manpower than if a transfer station is used. Another disadvantage of this operation is that the crew of packers, usually two or three, remain idle while hauling the wastes to disposal sites and returning for collection.

Construction of transfer stations and utilization of trailer trucks could alleviate the problems. However, this requires capital expenditures that local governments cannot afford. When the management is done on a regional level, enough capital can be generated.

Some of the central cities of metropolitan areas, often, do not have adequate landfill capacity to dispose of their wastes. This necessitates transporting the wastes long distances. If management is done on a regional basis, the "core" cities could use the site capacities of peripheral counties of the metropolitan area.

If, with increasing emphasis on resource recovery and the scarcity of land for disposal in recent years, cities are turning toward incineration with power or steam generation, recovery of useful commodities from the wastes (recycling), utilization of wastes as a substitute for fuel in power plants, composting and other advanced technologies, then these plants cannot be operated by small local governments.

It should be noted here that although regional management has distinct advantages, the communities called on to participate are sometimes reluctant to have wastes of another part of the region disposed of in their area and are unwilling to host waste processing facilities. Substantial political problems associated with regional management may emerge, but the generation of technically or economically feasible waste management options and mixtures thereof can facilitate a publicly acceptable solution based on negotiable settlement or environmental mediation.

1.5 SCOPE OF WASTE MANAGEMENT OPERATIONS

Modeling solid waste management comprises the following operations:

- Collection of wastes from sources
- Transportation of wastes from the sources to facilities. The facilities may be transfer stations, incinerators, composting plants, or any of the several technologies to treat solid waste
- Processing wastes at these facilities
- Hauling processed wastes from the facilities to ultimate disposal sites
- Disposal at landfills

When the management is on a regional basis, the alternatives available to the planner in the design and operation of a system are large in number and the decisions to be made are many. The regional planner has at his option several technologies which a local manager does not have. The regional planner will have to make decisions on:

- Selection of locations of landfill sites
- Selection of waste treatment technologies for the particular region. Some of the technologies may be more suitable for a particular region than others. (For example, solid wastes can be used as a substitute fuel for coal-fired power plants, whereas that option may be unavailable because of lack of coal-fired plants.)
- The locations of processing facilities
- Capacities of these facilities
- Capacity expansion strategies for the facilities and timing
- Flow routing of the wastes from the wastes through processing facilities to ultimate disposal at landfills

The models developed should not be misjudged as rigidly determining an ultimate decision but rather be used as effective tools for efficient management. Before the planner makes decisions regarding political, social and other practical aspects of the problem which are difficult to quantify, the models will help choose the most economical technologically feasible system.

1.6 WASTE MANAGEMENT OPTIONS AND COSTS

Municipalities can choose from several alternatives for processing and disposing of collected solid wastes. These methods include disposal, preprocessing and disposal, or preprocessing, resource recovery, and disposal.

Sanitary landfilling is the most widespread method of waste disposal currently used, but sites are becoming increasingly scarce, thus, more land-saving methods have to be pursued. Processing and disposal, the second major alternative, involves shredding or incinerating the waste stream before disposal, thus reducing the volume of waste in the landfill. The third major alternative involves processing the waste stream, recovering and selling certain resources, and disposing of the remainder. This method of waste management is not common today. It does, however, offer the potential for recovering substantial portions of the materials in the waste stream.

The four basic options are: (1) sanitary landfill; (2) shredding facilities in combination with shredded waste landfills; (3) solid waste-derived fuel in combination with waste landfills; and (4) incinerator in combination with residue landfill. The model would, of course, include mixtures thereof, if technically feasible.

(1) *Sanitary landfill.* The landfill site is assumed to have a 20-year lifetime and be capable of holding 20,000 tons of waste per area. Twice as many areas as are actually needed for disposal are necessary to allow room for access roads and to isolate the site. Complementary capital equipment includes compactors, scrapers and truck-type tractors. Properly managed, the site can be reclaimed for other (sometimes recreational) uses after being terminated as a landfill.

(2) *Shredding facilities.* Shredding reduces the particle size and increases the density of solid wastes. When compared with unprocessed wastes, larger quantities of shredded wastes can be landfilled on the same amount of land. Shredding is also the first step in materials recovery.

(3) *Solid waste derived fuel.* While shredding can reduce landfill costs, the municipality choosing this option must still dispose of a large proportion of its solid wastes (between 80 and 90%, even if glass and metal fractions are recovered). One method of significantly reducing the amount of waste disposed is to recover the combustible fraction as energy. Two methods of recovery energy are built into the model: producing a solid waste-derived fuel, or generating steam by incineration.

(4) *Incineration.* Incineration is the second major form of energy recovery available to municipalities though at a price of inducing severe pollution problems by burning raw refuse in normal water wall incinerators. These pollution problems would be substantially reduced with a pro-

cessed solid waste derived fuel (see (3) and Schlottmann (1977). Two types of incinerators are available, depending on the capability requirements. Modular or package incinerators are used for processing less than 500 tons/day. Water wall incinerators are used for processing between 500 and 3000 tons/day. The combustion chambers are enclosed by closely spaced, water filled tubes that recover heat from the burning wastes. Steam is generated by built-in recovery boilers. One characteristic of these incineration techniques is that all materials recovery takes place prior to incineration. This is not the usual practice. Currently, materials recovery, if any, most frequently occurs after incineration by processing the residue.

Of the four basic options only two are proven: sanitary landfill and incineration. The others appear to be promising techniques that will supplement the options available for solid waste management.

The major objective of this book is to develop a technology choice model that could serve to generate test bed solutions for planning regional waste management strategies. Specific problems to be addressed are:

- To make explicit the waste management options that are encountered by regional planning agencies
- To assess various waste treatment and disposal technologies, individually and in combination
- To produce a framework to effectively answer 'what if' questions, based on sensitivity analysis of the model
- To provide a rationale in comparing technology choices in the waste management area

The underlying models are applied to three regional waste management systems, two in the Federal Republic of Germany and one in the U.S.

1.7 CONTENTS

The book is formally organized in seven chapters.

Substantial problems of waste generation prediction, site selection, facility operation, vehicle routing and treatment technologies have been indicated in Chapter 1. The second chapter contains an extensive historical survey of the available international work on the applications of Operations Research, Management Science Techniques and Systems Analysis to Solid Waste Management, to show major trends of approaches. Such models are primarily optimization models of cost analysis and facility siting.

The third chapter deals with the developments of static models for a regional solid waste system. Three models are described and special purpose algorithms to solve each model are developed and tested successfully for three selected areas.

The fourth chapter contains detailed case studies for all three of the above models. They involve regional waste management options for the Boston metropolitan area, and for the Munich and Nuremberg/Fürth metropolitan areas in the State of Bavaria, F.R.G. The data and the solutions are given, and interesting environmental policy conclusions are obtained.

A multi-period planning model is developed in the fifth chapter and a heuristic solution procedure is described.

Chapter 6 suggests an econometric approach toward identifying demand, waste generation, supply sources, production and cost functions.

CHAPTER 2

A Review of Solid Waste Management Modelling

2.1 DIFFICULTY OF MODELLING OF SOLID WASTE MANAGEMENT

There are a number of properties of solid wastes which compound the difficulties of decision-making and modelling. The term 'solid wastes' encompasses an almost incredible variety of materials. Ordinary household wastes, which themselves contain many different components, make up only a small part of the problem. Bulky domestic wastes, such as refrigerators, bedsprings etc., require a different mode of handling. Industrial waste materials in various forms present yet a different set of difficulties.

Historically, action has been taken when solid wastes have been viewed as causing community or regional problems, thus, garbage collection was undertaken as a governmental (community-type) activity out of concern for public health and aesthetic deterioration of neighborhoods, while the unseen accumulation of waste material from an industry has often been left as a problem to be handled by the individual firm.

The local nature of the problem has naturally led to a wide variety of local solutions. Each individual producer of waste, each firm, each municipality has developed its own method of handling its own particular problems. Only in the last twenty years, in the most advanced industrial countries have we witnessed a stronger drive to work together for a common solution on a national and even international scale (see OECD Report (1981)).

Two basic restrictions are reported that frequently make the working out of a unified solution on a regional basis quite difficult.

First, large existing capital investments in fixed facilities in each locality make the developments of regional solutions, which would be advantageous from a design oriented viewpoint, often uneconomical—at least up to the point of capital depreciation or technological obsoleteness.

Second, models and techniques which are useful in improving the situation in one locality may be quite useless in another place, because of differences within existing systems.

11

For purposes of systematization it is convenient to distinguish between at least two types of solid waste decisions.

One might be termed the long range policy decision. Included in this class are decisions relating to national resource management policy, e.g., the extent to which recycling and reclamation should play a part in solutions of solid waste problems, and the methods to be used to encourage and implement such policies.

The remaining set of decisions could be termed 'management decisions' in a narrower sense. They are more commonly encountered on local, regional and occasionally state levels and they are most frequently concerned with the planning and operation of facilities and equipment within an existing institutional and policy framework. This implies that, for the most part, we take as given the generation of solid waste in its current or projected varieties and quantities, and consider only the system used for collecting, transporting, storing, treating and disposing of this waste.

By proposing a regional solid waste management system we are looking at the design level at various systems that facilitate regional waste management, but we also take care of the operational aspects of such a system.

To this end, in general, a solid waste system consists essentially of four components: (1) the waste itself; (2) the fixed (or designed) facilities used for transfer, treatment and disposal; (3) the vehicles used for transporting the waste; and (4) the crew assigned to operate the system.

The research papers in the area can be classified into three categories: (1) mathematical models which use mathematical programming and other advanced operations research tools; (2) computer simulation models; and (3) other works: regression analysis, data collection and policy-analytical studies of general nature.

2.2 MATHEMATICAL MODELS

The works of Altman et al. (1971) through Wahi and Petersen (1972) deal with mathematical models. The mathematical models can be further classified into two categories: (1) models for the collection facet of waste management; and (2) models for selection of transfer stations, incinerators and other intermediate processing plants, and disposal sites for regional management.

Collection Models

Clark and Helmes (1970) deal with the problem of locating garages from which the collection trucks leave for the day's work and to which they return after the day's work. The authors argue that since 80% of total expenses for solid waste management are incurred in collection, a study of the problem of locating garages would be useful. The relevant costs are: (1) fixed costs of constructing and maintaining a garage, (2) variable cost of a garage depending on the number of trucks assigned to the garage and (3) transportation cost from the garage to the collection district at the beginning of the day and the transportation cost from the landfill site to the garage at the end of the day. The problem is formulated as a fixed charge problem and a heuristic algorithm by Male (1973) is used.

Kirby (1971b) rightly points out in a discussion that the number of trucks assigned to a garage must be an integer and not a continuous variable as the authors had assumed and goes on to suggest an integer programming formulation of the same problem.

In another paper, Clark and Helms (1972), point out that, when vehicles are chosen for inclusion in the collection fleet, little consideration is given to their suitability for satisfying varied requirements of collection districts at a minimum cost. They suggest that costs to municipalities can be reduced by maintaining various-sized packer trucks and develop a linear programming approach for selecting proper fleet size and type of packer vehicle.

Dantzig and Ramser (1959) deal with the problem of dispatching a fleet of trucks from a central depot to many demand points and minimizing the total distance travelled by the fleet, satisfying demands at all points. The trucks are all assumed to have the same capacity. They suggest a procedure based on a linear programming formulation for obtaining a near optimal solution. Clarke and Wright (1964) consider the situation wherein the trucks have varying capacities and develop an interactive procedure that enables rapid selection of an optimum or near optimum solutions. Pierce (1969) gives direct search algorithms for truck dispatching problems and Hausman and Gilmour (1967) consider a multiperiod truck delivery problem. It is clear that these studies can be directly used in assigning collection vehicles from central depots to different parts of the region to collect wastes, some problems of this type are treated by Bellmore et al. (1972).

Fuertes, et al. (1972), consider the problem of assigning trucks to collection districts and routing the trucks through collection districts. Emphasis is placed on developing procedures which have small data requirements and which are easy to implement. A political districting algorithm is used to generate compact and equitable daily districts for each truck. The routing of a truck is posed as a Chinese postman problem and solved using simple manual algorithms. The solutions obtained are not necessarily optimal. The authors include examples relating to the operations in the Boston area in the report.

Altmann et al. (1971), present a nonlinear programming model for assignment of crew for household refuse collection.

Regional Models

Anderson and Nigam (1967) considered the problem of transporting wastes from transfer stations to landfill sites at a minimum cost, given that wastes should be shipped in aggregate amounts, such as in large transfer vehicles, and also that the aggregate flow is not too small relative to the total flow leaving a source. They developed a branch and bound scheme to solve the problem, but suggest that the procedure be abandoned if the number of landfills is greater than three in favor of a linear programming method.

The authors also developed an in-kilter algorithm to solve the problem of transporting wastes from the sources to landfills either through already existing intermediate processing facilities or directly, at a minimum total cost. Linear transportation and processing costs are assumed and capacity restrictions are placed on the flows through the intermediate facilities and landfills. The problem is different from a trans-shipment problem, because of the fact that flows through some of the nodes of transportation networks are not conserved. If, for example, 100 tons of wastes are sent into an incinerator plant, only 20 tons of ashes may come out of the plant.

The algorithm suggested is incomplete, fails to find flow-augmenting paths and hence incorrect—because the authors fail to recognize the existence of flow absorbing structures in the network. Jewell (1962), has developed a primal-dual method to solve a more general class of problems of minimizing the cost of flow through a network with 'gains.' The 'gains' are not restricted to be between 0 and 1 as is the case in the problem considered by Anderson and Nigam.

Helms and Clark (1971) present a mathematical model that will aid in selecting from among various alternative systems for waste management. Incinerators and landfills are considered as potential facilities for the system and it is assumed that the facilities have a fixed cost and a linear processing cost.

A mathematical program with linear constraints and a nonlinear objective function with fixed costs associated with 0–1 integer variables and linear processing and transportation costs associated with continuous variables is developed. The authors make an assumption that residues from incinerators would be taken to a landfill chosen a priori.

Marks and Liebman (1970) consider the problem of selecting transfer stations. The transfer stations have a fixed cost and a linear processing cost and wastes from the sources may be routed, through transfer stations or directly, to landfills. The objective of the model is to minimize the total cost of transportation, processing and fixed costs. A branch and bound scheme, in which the subproblem at each node of the branching tree are solved by Fulkerson's out-of-kilter algorithm (1962), is developed.

Rossman (1971) extended the work of Marks and Liebman (1970) by adding incinerators to the set of potential facilities. He used the branch and bound method developed in (1970) and employed Anderson's algorithm (1967) to solve subproblems at the nodes of branching tree.

Esmaili (1972) presents an optimization model to choose the combination of processing or disposal facilities, or both, from among a number of alternative facilities that would minimize the overall cost of haul, processing or disposal, or both, of solid waste management operations over an extended period of time. In summary, he concludes that the model also enables a comparative evaluation of the various methods of solid waste transport and provides a time-phased schedule for the required acquisition or construction activities, and the phasing out of operational facilities commensurate with the characteristics of facilities selected for the management system.

The facilities are assumed to have a fixed cost, a variable fixed cost and a variable operating cost depending on the loadings. In the model, if a facility was prematurely closed down, a capital cost discount factor (the amount of capital investment wasted) is used. The value of capital cost discount factor is a function of total capacity, age and operating capacity of the given facility. The processing plants are assumed to be capable of expansion without limits while the landfills have an ultimate filling capacity.

A heuristic solution procedure, which solves each time period problem heuristically and combines them to get a better solution, is developed. Little advanced mathematical programming technique is used in the solution procedure.

A long range planning model is presented by Fuertes et al. (1972). The problem considered is one of minimizing the total discounted costs, of transportation and processing, fixed costs and costs of expansion of facilities, over a period of time. Political and equity aspects of regional system operation are included in the model. The model was programmed on MPSX (Mathematical Programming System Extended), a "packaged" program product of IBM. Only a locally optimum solution could be obtained. One of the shortcomings of the model is that the cost of capacity expansion of a facility is assumed to be independent of its present capacity.

Nigam (1969) deals with the problem of finding the optimal size of plants to be built and their optimal installation times to minimize the present value of all future costs incurred, given a certain increasing demand function for a product and costs of operating processing plants of different sizes.

2.3 COMPUTER SIMULATION MODELS

The work by Berthouex and Brown (1969) through those of Clark et al. (1971) deal with computer simulation models of waste collection systems. An important difference between the computer simulation models and most of the mathematical models described earlier is that, in simulation models, the stochastic nature of amounts of wastes generated, efficiency of collection crew, speed of trucks etc., are taken into consideration while designing and analyzing a collection system, whereas average characteristics are assumed in mathematical models. Simulation models require statistical information on the factors under consideration.

Quon, Charnes and Wersan (1965) developed a model to simulate the operational characteristics of a refuse collection system using daily route method. The effect of queueing of trucks at transfer stations and disposal sites is taken into account and the model yields information on truck capacity utilization, collection efficiency, haul efficiency etc., as a function of the rates of waste generation, assignment of collection units, haul distance etc. In general, the model is suited for studying dynamic operational of existing or proposed refuse collection systems using the daily route method of collection.

Quon, Tanaka and Wersan (1969) report that the purpose of their model is the comparison of several operational policies of refuse collection, such as allowance of overtime, relay of the refuse collection vehicle between the garage and the disposal site by a driver after the regular working hours for the last load of the day, different assignments of refuse quantity expected and average length of workday for each collections crew, different number of trips to be made per day and varied frequencies of service to be provided.

In (1970), Truitt et al. have developed three simulation models, the principal purpose of which, the authors state, is to determine system responses as measured in cost per ton of collected refuse to various changes in operating policies. They investigate the need for a transfer station and costs of increasing collection frequency from twice a week to three times a week in some city (Baltimore) and claim, the models are general enough to be used in any other collection system. They conclude that a transfer station becomes economically justified at a haul distance of eight miles, regardless of collection frequency. Plots are made from simulation runs for cost per ton of refuse collected against haul distance, with or without transfer stations, with a collection frequency of two or three times a week.

Bodner et al. (1970) present a model to optimize collection routes of individual trucks on a daily or weekly basis. The model requires description of service area, collection equipment, waste generation rates and operational policies of the management. Since this is a simulation model, optimal routes are not guaranteed. Many feasible routes are generated randomly and the best of these routes is chosen.

Quon, Mertens and Tanaka (1970) wrote a simulation program to examine the relationship between the efficiency of refuse collection crews and: (1) the physical and population characteristics of the area served; (2) the type and condition of the equipment used; and (3) the several modes of operation. The efficiency was measured by dollars per ton of refuse collected, dollars per living unit serviced per week and loading speed. The conclusion reached on studying three wards of Chicago using the model, was that the age of truck and the rank (priority) of load strongly influenced work efficiency while the number of loaders on the truck and the population density had little effect on the efficiency of collecting crews.

Esmaili (1971), (1972) developed two simulation models. The first one can be used to select an optimal disposal site location in a given area, to minimize the total cost of waste transportation and disposal. The second model can be used to determine the effect of alternate facility locations and transport truck capacities on the total combined cost of transport and disposal operations.

Truitt, Liebman and Kruse (1961) built a simulator to investigate system response to changes in operating policy measured in cost per ton of collector refuse. They looked at two policy areas: the use of transfer stations and increases in collection frequencies. They simulated trucks moving from a disposal site to a collection area, through that area and back to disposal. No attempt was made at routing the vehicle through the area, and collection times were based on average collection speeds and times for a given waste load to be collected, all of which were generated randomly, from observed system histograms. Since they planned to look at the Baltimore System, the basic work-rule policy was to assign given routes rather than a constant length workday. Provision in the model for the location of one given transfer site was also included.

The data needed for the simulation were these:

- Boundaries of the area, and of the subareas with their population-density classifications
- Frequency of collection
- Collection-truck capacity
- Transfer station location (if any)
- Size and type of transfer trailers (if any)
- Location of final disposal site
- Labor pay scales, including overtime rate

- Probability distribution functions for collection rates, unloading times, truck speeds and amount of waste generated per capita per day
- Average number of persons per housing unit for each neighborhood density classification
- Amortization interest rate and time period for transfer-station structures and equipment
- Capital investment in transfer-station land
- Hourly operating cost for tractor-trailer rigs as in the previous item
- Monthly cost of utilities for the transfer stations

Each simulation run was for one week, and the program calculated the number of trucks needed and assigned areas to each of these trucks for each day. The northwest quadrant of the city of Baltimore, Md., with 15,000 housing units and 225,000 persons, was chosen for study with the model. After verification runs, a set of experiments were run to show the cost of increasing collection frequency based on the assumption that collection time per ton would increase due to decreased loadings at each stop but that the total waste load collected would not change with frequency. The resulting increase in collection costs were about 9% with 60% of that amount due to equipment increases and 40% to labor. The investigators also noted that increasing collection frequency also increases the peak loads to the disposal facility. Investigation of the possible use of a given transfer station site showed that at the present haul distance the cost savings are marginal. A greater distance to the disposal point was necessary to justify building and operating a transfer facility.

2.4 ROUTE GENERATION MODELS

This class of models addresses the basic question of how individual routes and assignments for crews may be specified. The problem of vehicle routing is a fairly general problem in management and has received considerable attention in the past few years because of the obvious payoffs to be gained in increasing routing efficiency. However, the very large scale needed to deal with the problem in the context of an entire city involving thousands of streets immediately makes the task of generating good routes quite challenging.

The problem is characterized by the fact that it is not too difficult to build a feasible route or set of routes which satisfy all necessary time, distance and volume constraints. However, generating very good or even optimal routes is extremely difficult because the number of feasible routes increases exponentially with the number of streets. Attempts at generating both good feasible routes (heuristic methods) and optimal routes will be presented with emphasis on those that have been applied to solid-waste problems. One way of approaching the problem is to break areas into subareas involving very small collection tasks and then look at good ways of combining them into daily or weekly crew assignments.

Sample problems based on large cities indicate substantial savings of millions of dollars per year due to the efficiencies gained. It would seem that in view of the problems encountered in actual routing in large cities that heuristic approaches will yield substantial benefits. Bodner, Cassell and Andros (1970) have built a simulation model which operates at the scale of the street network and generates feasible routes randomly. They consider such parameters as crew size, vehicle size and pickup time. Required for the simulation is a detailed representation of the street network which could severely limit a large problem. Waste loads and collection times are generated from distributions and the truck is routed by making a decision at the end of each street of whether to go to disposal because of a capacity restriction, to the garage because of a time constraint or randomly to choose another street in which to collect. With repeated application of the techniques a large number of feasible routes are generated from which the "best," based in their case on minimum distance travelled, is chosen. It should be noted that other criteria, such as minimum time, for good routes might also be applicable. They chose as an example a small town (Potsdam, N.Y.) which has a population of 11,000 and uses one truck for once-a-week collection. In the example, 100 runs were made with no stochastic variation of load and the minimum distance route was chosen. There was a significant difference between the average route distance for the trials and the minimum distance route selected. The investigators refer to this as the "optimum" route which is not necessarily true since only a few alternatives have been searched. However, in a statistical sense, this size sample should produce a fairly good feasible route. The chosen route was then subjected to a long-term simulation with stochastically varying waste loads.

Looking at the general problem of routing vehicles, Marks and Liebman (1970) presented a detailed description of the various techniques attempted in the literature. Basically vehicle routing problems may be categorized by the responses to the following quotations:

(1) Are the services to be rendered by the vehicles located at a finite number of points which have given location coordinates (the discrete problem) or are they in essence continuous in that every street or set of streets must be covered (the continuous problem)?

(2) Is vehicle capacity (defined both in terms of volume and weight) such that the vehicle can accomplish the entire servicing operation in one continuous route? If so, the problem is a single-route problem. If not, either more than one vehicle is needed or the single vehicle must return to its origin (such as the garage or disposal points) more than once. In either case, we are dealing with a multistate problem.

(3) Are there additional constraints on routing such as precedence, earliest and latest arrive time, etc.?

Travelling-Salesman Problem

The case of single-route discrete problems without additional constraints is known as the "travelling-salesman" problem. Here the objective is to find the minimum-distance tour a salesman would follow to visit N cities (or points) and return to his origin so that all cities are visited exactly once and his tour is continuous. Distances between all city points are assumed known and may or may not be symmetric. For the asymmetric case, an N-city problem has (N-I)! feasible solutions, a number which very rapidly becomes large as N increases. Success in recent years has been limited to problems involving 100 cities or points which are not large when applied to solid-waste collection. However, in this case we are dealing with multi-routes. This problem has been called the M-salesman travelling-salesman problem by Marks and Liebman (1970) and computational success with any fair-sized problem has been extremely limited. Most work in this area deals with heuristic methods of establishing good feasible routes (B. Korte, 1983).

Chinese-Postman Problem

One would certainly argue that viewing the solid-waste collection routing problem as discrete rather than continuous may be greatly enlarging the problem. Until recently, the only way to ensure this has been that at least one could approach a solution of the discrete problem but had no way of dealing with the continuous problem. However, recent developments not only show us a hint of how to deal with continuous problems but indicate the ability to deal with much larger problems in a shorter time.

The single-route continuous problem is called the "Chinese-postman" problem. It is defined as finding the minimum distance continuous tour through a network which travels all arcs at least once. Thus the Chinese-postman problem is an arc-covering problem while the travelling salesman problem is a node-covering problem. Recent developments in the Chinese-postman problem using graph theory point to quick and efficient methods of solution.

2.5 OTHER STUDIES

Works by Clark and Gillean (1975) to Midwest Research Institute (1979) deal with various aspects not dealt with by models in previous groups. Two of these papers will be described here.

Clark et al. (1971) suggest a regression analysis for arriving at the average cost of metropolitan wide solid waste management, as well as the factors influencing the cost. They considered the following variables for the analysis:

Y = average annual budgeted cost of refuse collection in \$
X_1 = yearly collection frequency
X_2 = combined or separate pickup
X_3 = pickup location, i.e., from the curb or at the rear of the house
X_4 = crew size
X_5 = pickup density, i.e., the number of residential pickup units per square mile
X_6 = nature of financial arrangement.

There are at least four major contractual arrangements by which solid waste is collected: (1) city operations; (2) city contract; (3) city license; and (4) private arrangements. The working hypothesis was formulated as:

$$Y = f(X_1, X_2, X_3, X_4, X_5, X_6).$$

A stepwise regression was used to examine collection cost data for 20 municipalities in a metropolitan area in Ohio, i.e., a simple regression relation between y and x was estimated by inclusion of additional regression variables.

The analysis concludes that financial arrangement, collection frequency and pickup location are the only significant factors.

DeGeare et al. (1971) conducted a study to determine the variables that best describe the quantities of solid wastes operated by selected types of commercial enterprises. Data were collected during a three-week field study in the greater metropolitan area of Cincinnati, Ohio. The variables considered for the study were:

Y = weekly quantities of solid wastes collected
X_1 = number of hours open per week
X_2 = number of business days open per week
X_3 = average annual gross receipts

X_4 = physical area of the store in square feet
X_5 = average inventory in dollars
X_6 = equipment value in dollars
X_7 = number of delivery days per week and
X_8 = number of employees.

The types of retail establishments selected for the study were clothing, drug, grocery and hardware stores and restaurants. A stepwise regression analysis was carried out to relate Y to the variables X_1 and X_8.

The conclusion of the study was that the generation of commercial solid wastes closely related to the number of employees, hours open, and type of establishment involved. The authors believe that this preliminary study can be helpful to those interested in commercial solid waste generation studies and to planners and engineers of solid waste handling systems.

Other papers, Gouleke (1970), Midwest Research Institute (1979), discuss at length waste generation prediction and facility problems.

CHAPTER 3

A Model of Waste Management

3.1 INTRODUCTION

The total system of waste management can almost be decoupled into two major subsystems. One is the solid waste collection system and the other is the regional management system. The assumption of decoupling is made for conceptual purposes. Frequently there seem to be interactions among both systems and under these circumstances a waste management solution may be vastly improved by modelling both systems as one (Ward and Wendell, 1978). A solid waste collection system is concerned with the collection of the wastes from the sources, routing for trucks within the region, the frequency of collection crew size, truck sizes, number of operating trucks, transportation of collected wastes to a transfer station, an intermediate processing facility or a landfill and a host of other problems. The regional management concerns itself with the selection of the number and locations of transfer stations, intermediate processing facilities, landfill sites, their capacities, capacity expansion strategies and routing of the wastes through the facilities to ultimate disposal on a microscopic level. While the wastes are generated all over the region, it is assumed that they are generated at point sources, each point source representing a neighborhood. The routing of wastes from these point sources to ultimate disposal is considered to be a part of regional management.

The general problem of the (static) allocation model of the waste management system can be described as follows: Given the potential locations of intermediate processing facilities and landfills, the locations and capacities of existing facilities, the cost structures—transportation, processing and fixed costs—and the quantities of wastes generated at the sources, find out which facilities should be built and how the wastes should be routed processed and disposed of so that the overall cost of the system is minimized.

This problem is actually equivalent to a large class of problems pertaining to optimal network design (Handler and Mirchandani, 1980; Trietsch, 1985). A broad class of models for regional waste management has been developed over the past 20 years, as reviewed and summarized in the previous chapter. The class of models that we set out to describe next is of the fixed charge mixed integer programming type, which views regional waste management system as network flows. More technically, we point out an affinity of our model with that of Walker (1974), with our algorithm being an extension and improvement over that of Marks and Liebman (1970). Related and more recent work of this kind includes Clark and Gillean (1975), the WARP Model, Hasit and Warner (1981), and the Resource Recovery Planning Model, Chapman and Yacowitz (1984).

Before developing the models for the system, a brief description of potential components of the system are given.

Considerable emphasis has been placed on resource recovery from solid wastes in the recent years. The Midwest Research Institute in its report [1973], to the Council of Environmental Quality (U.S.) classifies the various resource recovery processes into the following general categories:

(1) *Energy recovery processes*. Processes that recover the energy content of mixed municipal wastes, in the form of either steam, electricity or fuel.

(2) *Materials recovery processes*. Processes which separate and recover the basic materials from mixed municipal wastes, such as paper, metals and glass.

(3) *Pyrolysis processes*. Processes that thermally decompose the waste in controlled amounts of oxygen and produce products such as oil, gas, tar, acetone and char.

(4) *Compost processes*. Processes which produce a humus material from the organic portion of the mixed waste.

(5) *Chemical conversion processes*. Processes which chemically convert the waste into protein and other organic parts.

The economics of the processes are displayed in Table 3.1, reproduced from the above mentioned report.

In modelling a regional system the planner will have to choose the processes that are applicable to the region. For example, although fuel recovery process wherein mixed waste is used as a fuel in power plants, is the most economical process, it requires that a power plant that could use wastes as fuel exists in the region. Similarly, a composting process requires that there exist markets for the products of the process.

Once the decision regarding the selection of the processes is made, the planner will have to choose potential locations for the facilities in the region. It should be noted here that this is a very important and difficult aspect of the problem. There are many social and political hurdles to cross before the planner can decide on potential locations for the facilities. Although many

Table 3.1. Summary of resource process economics[a]

Process Concept	Investment ($000)	Total Annual Cost ($000)	Resource Value ($000)	Net Annual Cost ($000)	Net Cost p. Input Ton ($)
Incineration Only	9,299	2,303	0	2,303	7.68
Incineration and Residue Recovery	10,676	2,689	535	2,154	7.18
Incineration and Steam Recovery	11,607	3,116	1,000	2,116	7.05
Incineration and Steam and Residue Recovery	12,784	3,508	1,535	1,973	6.57
Incineration and Electrical Energy Recovery	17,717	3,892	1,200	2,692	8.97
Pyrolysis	12,334	3,287	1,661	1,626	5.42
Composting (Mechanical)	17,100	2,987	1,103	1,884	6.28
Materials Recovery	11,568	2,759	1,328	1,431	4.77
Fuel Recovery	7,577	1,731	920	811	2.70
Sanitary Landfill (Close-in)	2,472	770	0	770	2.57
Sanitary Landfill Remote	2,817	1,781	0	1,781	5.94

[a]Based on municipally owned 1000-TPD plant with 20-year economic life, operating 300 days/year.
Source: Midwest Research Institute.

communities would like to reap the benefits of regional management they are generally not willing to play the host to processing facilities.

We can now describe and justify the two classes of models. The models of the first class are called the static models and of the other class dynamic model. For various explanations of dynamic models we refer to the survey of Sethi and Bookbinder [1980]. In the static models, the wastes generated in the region are assumed to remain at a constant level throughout the planning horizon.

On the other hand, there are metropolitan areas where waste generation is increasing every year and the dynamic model described in Chapter Five, appropriately deals with such a situation. Static models can also be used by planners who plan for a short period of time, less than say 5–8 years, and believe that for that period the assumption of constant waste generation is a reasonable approximation to the actual situation.

3.2 COST STRUCTURE

If we choose to use some of our resources to develop and produce a certain capability, then those resources are obviously not available for the production of some other, perhaps superior, capability. An estimate of the cost of any such choices or decisions is an estimate of the benefits that could otherwise have been obtained. "Economic costs" are benefits lost. For this reason economic costs are often referred to as "opportunity costs." It is in alternatives, in *foregone opportunities,* that the real meaning of "costs" must always be found.

The two types of cost components have to be distinguished conceptually: *transportation costs* and *facility costs.*

The cost of transportation of wastes from the sources to intermediate facilities or landfills, and from intermediate facilities to landfills depend on: the distance traversed; the amount of wastes transported; type of waste transported; number of crew men; type of transportation used, 'Packers' or trailer trucks; frequency of collection; routing of trucks; and payscales of the workers; among the other factors.

Since we are dealing with regional management, and since including all these factors in the model explicit makes the model computationally intractable, we assume a cost which is linearly proportional to the distance travelled and the amount of wastes transported. The other factors are not explicitly considered. The cost will also be assumed to depend on the type of waste transported—industrial waste, municipal waste, demolition debris, etc. The shortest distance between two points in the region "L" is used to calculate the transportation costs.

The capacity on the routes between two points in the region is assumed to be boundless. If, for some social or political reasons, wastes from point A cannot be transported to point B, the cost of transportation between the two points is assumed to be very high, making the route uneconomical in the final solution.

In estimating facility costs, we adopt the following approach. For annual costs the operations of each facility includes:

(1) *Direct operating costs.* Labor, materials, supplies, utilities, variable overheads and other expenses related directly to the amount of waste material handled;
(2) *Fixed costs.* Those incurred with time rather than with the scale of plant operation, and covering the costs of maintaining and administering overall operation; and
(3) *Capital charges.* Those directly attributable to the amount of capital tied up in the facility. These capital charges, in turn, cover capital recovery and amortized investment (depreciation and interest); insurance and administrative charges applied to fixed plant; and interest on recoverable investment.

Various types of cost functions have been discussed explicitly by Walker et al. (1974), i.e., linear, piecewise linear and non-linear facility costs. The subsequent analysis is related to this discussion.

There are two different cases we have to consider, while discussing the facility costs for our models. One is the case when a facility already exists in the region and has a finite capacity. It is assumed in the models developed here that such a facility has a fixed cost independent of flow through the facility and a linear processing cost. The cost function of the facility is shown in Fig. 3.1.

In the second case, the facility does not exist, but needs to be designed. The best strategy would be to build a facility of capacity equal to the flow through the facility, when we are dealing with static models. This is based on the assumption that once the facility is built, the flow through it will be sufficiently large that any excess capacity would be uneconomical.

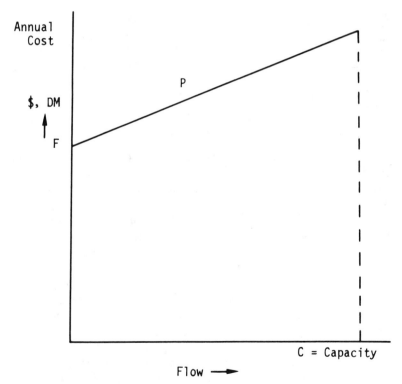

Figure 3.1. Cost function of an existing facility.

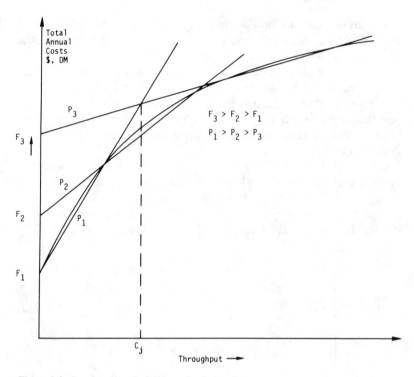

Figure 3.2. Cost function of a facility to be designed.

Since costs are estimated as explained above, and the total annual costs are plotted against the flow (or capacity, since the design would be to build a capacity equal to the throughput), the curve will look like the one in Fig. 3.2.

The cost function is now approximated by linear segments as shown and the facility is assumed to be equivalent to three (n in a general case) 'pseudofacilities,' each with a fixed charge and a linear processing cost. A 0–1 integer variable is associated with each pseudofacility with the condition that only one of them may be built. Actual plots are given in Chapter 4.

It should be borne in mind that the empirical interpretation of costs in the analysis of economic models for solid waste management is that of 'opportunity costs' for alternative uses of resources. Thus, we are concerned with evaluating alternatives, in particular, when we choose limited resources to allocate waste among various disposal modes then those resources are not available for other disposal modes where they could produce perhaps superior benefits. Thus, strictly speaking, we define the economic costs of a disposal or management system as the value of benefits that might have been obtained through alternative uses of the resources within that system. Although we maintain that the real meaning of cost is to be found in the benefits *foregone,* in deciding to apply resources in one way rather than another, it is usually the case that costs are measured not in terms of benefits foregone but in terms of money expenditure required to implement the program of concern.

The great analytical advantage of the models proposed for solid waste management is that this 'opportunity cost principle' is fully built into the framework and is clearly reflected by the practical results obtained through the case studies in Chapter 4.

3.3 MATHEMATICAL MODELS

This section deals with the application of a group of mathematical models to the systematic analysis of the solid waste management problem. The purpose of such models is twofold: to model the waste management system in order to show its sensitivity to various parameters, and to predict the outcome of possible policy changes so that alternative management schemes may be evaluated.

Ideally, the analyst could build a simple model of the solid waste management system and learn quickly all there is to know about its properties and optimal control. This is not the case here. It should be emphasized here that a combination of a complex, interactive, technical, social, environmental, political and economic system to be approached, and the relative simplicity of the models available to the analyst tends to make the approaches suggested here more first-cut screening, than for making explorators choices as tools for detailed design and planning. To answer some of the logistics and allocations problems of solid waste management, the analyst would like to build a model that would be able to map the waste management process in its smallest detail and be able to manipulate all of the myriad parameters that could possibly affect the solution. However, its magnitude and complexity are such that to consider the system in its entirety and to encompass every possible detail is quite difficult, if not impossible. For this reason, simplicity assumptions must be made in model development which will allow some approximation but not the firm details of the problem. In the vehicle routing and waste collection problem, basically two different approaches are taken. If one is willing to neglect the problems of routing of the individual vehicles through their collection tasks, the large-scale problem of how material should be moved through the system and how transfer-facility location may be chosen can be approached optimally. If the location of facilities and collection areas assigned are known for a set of collection vehicles, the small scale problems of routing vehicles among individual tasks may be approached.

A mathematical formulation of the problem is given below.

General Mathematical Model

PROBLEM A.

$$\min \sum_{j \in I \cup L} \sum_{i \in S} f_{ij} t_{ij} + \sum_{k \in L} \sum_{j \in I} f_{jk} t_{jk}$$
$$\tag{1}$$

$$+ \sum_{j \in I \cup L} p_j \sum_i f_{ij} + \sum_{j \in I \cup L} F_j y_j$$

subject to

$$\sum_{j \in I \cup L} f_{ij} = G_i, \quad i \in S, \tag{2}$$

$$\sum_{i \in S} f_{ij} - a_j \sum_{k \in L} f_{jk} = 0, \quad j \in I, \tag{3}$$

$$\sum_{i \in S \cup I} f_{ik} \le C_k y_k, \quad k \in L, \tag{4}$$

$$\sum_{i \in S} f_{ij} \le c_j y_j, \quad j \in I, \tag{5}$$

$$\sum_{j \in B_i} y_j \le 1, \quad i \in F, \tag{6}$$

$$f_{ij} \ge 0 \tag{7}$$

where

f_{ij} = flow from i to j
y_j = 0 or 1 if j $\in$ B$_i$ for i $\in$ F
y_j = 1 if the facility j is already in existence and
S = set of sources
G_i = quantity of waste generated at source i

I = set of intermediate pseudofacilities yet to be built and the intermediate facilities already in existence

L = set of pseudofacilities at landfill sites that may be 'built' and the landfill facilities that already exist

F = locations of potential facilities

B_i = approximation to a facility cost curve at location i

t_{ij} = transportation cost from i to j

p_j = processing cost at (pseudo) facility j

F_j = fixed cost at (pseudo) facility j

C_k = capacity of (pseudo) facility k.

The first two terms of the objective function (1) give the total transportation costs; the third term gives the costs of processing at facilities and the fourth term gives the total fixed cost incurred.

Equation (2) implies that all the wastes generated at all sources should be shipped out. Equation (3) is a balance equation between the input and output at intermediate facilities. Equation (4) implies that at no facilities the amount of wastes treated can be greater than its capacity. At any potential facility location, only one pseudofacility can be built and this condition is reflected in inequalities (5) and (6).

When represented in a graphical form, the *flow network* looks like the one shown in Fig. 3.3. There are three sets of nodes s, the sources, I the intermediate facilities including the ones to be designed and the ones that already exist and L the landfill sites including the ones to be 'built' and the ones that already exist. There are arcs between s and I, s and L and I and L; there are no arcs between the nodes of the same set. The nodes of I and L have capacity restrictions, costs and zero-one integer variables associated with them. In addition, the flow is not necessarily conserved at intermediate nodes I. In other words, if the input to node i ϵ I is x, the output from the node is $a_i \cdot$ x where $0 \leq a_i \leq 1$. For example, if 100 tons of wastes are sent to an incinerator, the output may be 20 tons of ashes.

To represent the capacities and costs at the nodes and to represent the condition that all the wastes generated in some source s ϵ S must be disposed of, the network is enlarged with a super source 0, a super sink t and additional arcs as shown in Fig. 3.4.

The arcs $(0, s_i)$, i ϵ S, have a lower bound of zero and an upper bound equal to the amount of wastes generated at source s_i. The costs on the arcs $(0, s_i)$, i ϵ S, are set to zero.

For each intermediate facility, three nodes and two arcs are introduced. The capacity and processing cost are introduced in the arc (i_1, i'_1) as shown in Fig. 3.4. The reduction factor a_i is introduced in another arc (i'_1, j''_1) as $k'_i i''_1$ the capacity of which is infinity and the cost is zero. This 1 will be the only

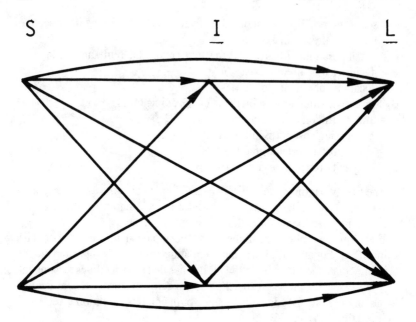

Figure 3.3.

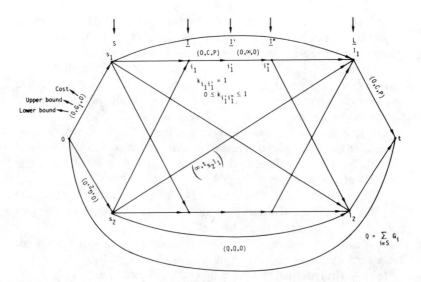

Figure 3.4.

type of arc which has a reduction associated with it. The advantage of having this additional arc will become apparent later on. The capacity and processing cost of a landfill are represented on the arc (lt) introduced in the enlarged network for each l ∈ L.

Having described the network we will be dealing with, the solution procedure for solving the mathematical program (1)–(7) can be outlined. The solution procedure is a branch and bound technique very similar to the one developed by Marks (1970), but special purpose algorithms are given to solve the subproblems of each node of the branching tree.

Branch and bound methods are exact procedures in the sense that they will produce optimal solutions. In a particular branch and bound application, however, there is no guarantee, a priori, that this approach will be significantly better than complete enumeration. In practical applications it may turn out that the entire branching tree cannot be exhibited due to storage or computational complexity type limitations. In such cases we discover the value of *heuristic* procedures. They provide a simple alternative to more complicated exact algorithms. Especially when optimal solutions are not absolutely necessary, the computational savings may be cost beneficial. But heuristics are equally important when used in conjunction with many exact algorithms. Besides providing a good initial feasible solution, a heuristic solution also provides an upper bound on the optimal objective function value which enables us to obtain faster convergence (to optimal solutions) in the mathematical programming algorithms. In turn, the optimality of a heuristic solution, or its maximum error, can be checked by the tight lower bounds obtained from the linear-programming based algorithms.

As stated earlier, the concave cost function of a potential facility is approximated by several straight lines. A single potential facility is approximated to be equivalent to several pseudofacilities with different fixed costs and linear processing costs. The number of pseudofacilities introduced in approximation to the one in reality is equal to the number of straight lines used to approximate the concave cost function. A 0–1 variable is introduced for each such pseudofacility with the condition that only one of the pseudofacilities may be built. This side condition is taken care of in the branching procedure.

It is clear that if the integer variables y_j's are known, then the mathematical program (1)–(7) is a minimum cost network flow problem and appropriate network algorithms can be developed to solve the problem. This special structure of the problem is exploited in the solution procedure for the problem (1)–(7).

The cost curve of a pseudofacility is shown in Fig. 3.5. At the beginning of the branch and bound procedure the cost functions of every pseudofacility is replaced by a modified cost MC given by MC = F/C + p as shown in Fig. 3.5.

The capacity of each pseudofacility is unbounded but it is taken to be equal to the total amount of all the wastes to be sent through the network from the sources. Since this is the maximum possible flow through the pseudofacility, we can, for algorithmic purposes, treat the capacity of the pseudofacility as infinite. The setting of the capacity of a pseudofacility will be discussed further as we deal with different types of static models. At the moment it suffices to say that the capacity of a pseudofacility should be set at the maximum of the total amount of wastes generated in the region. Setting any higher capacity would result in the decrease of modified cost and consequently, a reduced lower bound for the optimal solution to the problem at each node of the branching tree.

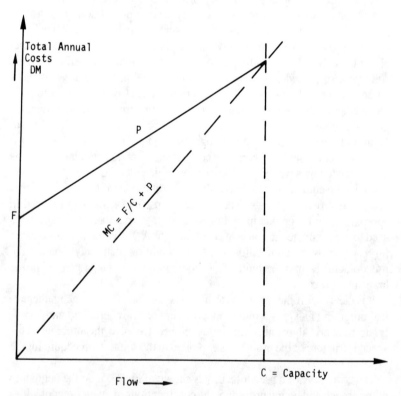

Figure 3.5. Cost function of a pseudofacility.

If, as shown in Fig. 3.2, the concave cost function of a facility is approximated by n (n = 3 in Fig. 3.2) linear segments, then the pseudofacility with the highest fixed cost and the least variable cost will have the lowest modified cost, and hence, if any flow is sent through a pseudofacility at a location, it would be sent through this pseudofacility. So the other pseudofacilities may be dropped from the network. This reduces the number of nodes and the number of arcs in the network. As will become clear, at any node of the branching tree, there will be only one pseudofacility at each location in the network.

With modified costs given to the pseudofacilities with the highest fixed cost and the least processing cost at each location and with the integer variables of these pseudofacilities set to 1 (the integer variables corresponding to other pseudofacilities are set to zero), Problem A can be restated as follows.

PROBLEM B.

$$\min \sum_{i \in S} \sum_{j \in I^* \cup L^*} f_{ij} t_{ij} + \sum_{k \in L^*} \sum_{j \in I^*} f_{jk} t_{jk}$$

$$+ \sum_{j \in I^* \cup L^*} F_j / C_j + \sum_{j \in I^* \cup L^*} p_j \sum_{i \in S \cup I^*} f_{ij} \tag{8}$$

subject to

$$\sum_{j \in I^* \cup L^*} f_{ij} = G_i, \quad i \in S, \tag{9}$$

$$\sum_{i \in S} f_{ij} - a_j \sum_{k \in L^*} f_{jk} = 0, \quad j \in I^*, \tag{10}$$

$$\sum_{i \in S \cup I^*} f_{ik} \le C_k, \quad k \in L^*, \tag{11}$$

$$\sum_{i \in S} f_{ij} \le C_j, \quad j \in I^*, \tag{12}$$

$$f_{ij} \ge 0 \tag{13}$$

where I* = set of pseudofacilities that have the highest fixed costs and the least processing costs at potential intermediate facility sites and the intermediate facilities already in existence.

L^* = the set of pseudofacilities that have the highest fixed costs and the least processing costs at potential landfill sites and the landfill sites already in existence.

Looking at the constraint sets of problems A and B, it is clear that every feasible solution of problem B is a feasible solution of problem A. The constraints of problem B are more 'relaxed' than that of problem A. In addition all the modified costs are lower than costs of problem A and hence the cost of an optimal solution to problem B is a lower bound on the cost of optimal solution to problem A.

If upon solving problem B, all the facility flows are either zero, or equal to the capacity, we have arrived at an "Exact Solution" and the optimal solution of problem B is an optimal solution of problem A. However, if the flow in a pseudofacility is less than the capacity but greater than zero, the cost of this solution is less than the cost of the optimal solution of problem A. We now branch on this facility. One at a time, each pseudofacility is assumed to be built and the problem is solved with its true linear processing cost and fixed cost. In addition, another problem in which it is assumed that there exists no facility at this location, is also solved.

All the solutions are placed on a master list and the least cost solution is taken. If the solution is such that, for all pseudofacilities whose y_j's are yet to be chosen, the flow through the pseudofacility is either zero or equal to the capacity (in other words, if the solution is exact), we have arrived at an optimal solution for problem A. Otherwise, we branch again on a pseudofacility which has a flow greater than zero but less than the capacity, as described above. The procedure terminates because all the different combinations of 0–1 integer variables will be formed by the branching routine.

At any general node of the branching tree, the mathematical program to be solved may be written as follows:

Define T as the set of integer variables that are set to 0 or 1 and T as the set of integer variables that are not yet decided upon. $T \cup T$ is the complete set of integer variables and $T \cap T$ is a null set.

PROBLEM C.

$$\min \sum_{i \in S} \sum_{j \in I \cup L} f_{ij} t_{ij} + \sum_{k \in L} \sum_{j \in I} f_{jk} t_{jk} + \sum_{j \in T \cup T} F_j y_y + \sum_{j \in I \cup L} p_j \sum_i f_{ij} \qquad (14)$$

subject to

$$\sum_{j \in I \cup L} f_{ij} = G_i, \qquad i \in S, \qquad (15)$$

$$\sum_{i \epsilon S} f_{ij} - a_j \sum_{k \epsilon L} f_{jk} = 0, \quad j \epsilon I, \tag{16}$$

$$\left. \begin{array}{ll} \sum_{i \epsilon S \cup I} \cdot f_{ik} \leq C_k & \text{if } y_k = 1, \\ = 0 & \text{if } y_k = 0, \end{array} \right\} k \epsilon (T \cap L), \tag{17}$$

$$\left. \begin{array}{ll} \sum_{i \epsilon S} f_{ij} \leq C_j & \text{if } y_j = 1, \\ = 0 & \text{if } y_j = 0, \end{array} \right\} j \epsilon (T \cap I), \tag{18}$$

$$\sum_{i \epsilon S \cup I} f_{ik} \leq C_k y_k, \quad k \epsilon T \cap L, \tag{19}$$

$$\sum_{i \epsilon S} f_{ij} \leq c_j y_j, \quad j \epsilon T \cap I, \tag{20}$$

$$\sum_{j \epsilon B_i} f_{ij} \leq 1, \quad i \epsilon F, \tag{21}$$

$$f_{ij} \geq 0. \tag{22}$$

Instead of solving problem C, a lower bound to the cost of the optimal solution to the problem can be obtained by solving a problem wherein the pseudofacilities for which y_j's are unknown are replaced by pseudofacilities with the least modified costs. The y_j's of other pseudofacilities are set to zero. This approach is similar to solving problem B instead of problem A.

3.4 GENERAL ALGORITHM

1. Input Data:

  m = number of locations of intermediate facilities and landfills.

  n = number of sources.

  B_i = number of straight lines used to approximate the concave cost function of a facility in site i, number of pseudofacilities in site i.

  m^* = total number of pseudofacilities = $\sum_{i=1}^{m} B_i$.

  F_j = fixed cost of pseudofacility j, \$.

  P_j = processing cost of pseudofacility j, \$/ton.

  F_{i*} = the highest fixed cost of all pseudofacilities located in site i = $\max_{j=1,...,B_i} \{F_j\}$.

P_{i*} = the lowest processing cost of all pseudofacilities in site i = $\min_{j=r_i,...,B_i} \{P_i\}$ $/ton.

$i*$ = the pseudofacility which has fixed cost = F_{i*} and processing cost P_{i*}.

t_{ij} = transportation cost from i to j (i ϵ S, j ϵ I)(i ϵ S, j ϵ L) and (i ϵ I, j ϵ L) $/ton.

S = set of sources.

T = set of intermediate facilities.

L = set of landfills.

2. Define T = set of integer variables y_j's set to zero or one.

      I = set of all integer variables which are not decided upon.

      T∪T = set of all integer variables.

      T∩T = ϕ'.

      T' = set of integer variables set to 1.

      T^0 = set of integer variables set to 0, T∪T^0 = T.

Initialize: T = set of all integer variables, corresponding to the pseudofacility i* for each location i, i = 1, . . . , m. T = complement of T, T = T^0∪T' = ϕ.

3. For all the pseudofacilities which belong to T, take the real processing cost and fixed cost. For all the pseudofacilities in T set the modified cost

$$MC_j = \left(\frac{F_j}{C_j} \right) p_j \quad \forall j \epsilon T$$

$$MC_j = p_j \quad \forall j \epsilon T'$$

and capacity $c_j = 0$, for all j ϵ T^0.

4. Use one of the special purpose algorithms described later in the chapter to solve the minimum cost flow network problem. If there is no feasible solution and this is the first time through, stop. No feasible solution exists for problem A. If it is not the first time through, no branching will be done from this node and hence the information about this need not be stored. Go to step 6 if this is the last of the problems set up in Step 7. Or else continue with the next problem defined in Step 7 and go to Step 3. If an optimal solution is obtained (the solution cannot be unbounded because the problem is to send a finite amount of flow through the network and even if the costs are negative the minimum cost of the solution cannot be infinitely negative).

Compute

$$\text{Cost} = \text{minimum cost of network problem} + \sum_{j \in T'} F_j.$$

Compute also the upper bound to the problem defined on the node as

$$\text{UPPER BOUND} = \text{COST} + \sum_{j \in T \cup G} F_j$$

where G is the set of facilities which have flows greater than zero.

Step 5.

Test the 'exactness' of the solution. A solution is 'exact,' if in all facilities belonging to the set T, the flow is either zero or equal to the capacity.

If the solution is not exact, then compute the ratio of the total flow into the facility to the capacity of the facility for all the facilities belonging to set T and having flow between zero and capacity. Choose the facility which has this ratio closest to .5. Store all the information about the problem. If this is the last of the problems defined in Step 7, go to Step 6. Or else, continue with the next problem defined in Step 7 and go to Step 3.

Step 6.

Select the problem which has the least cost from the storage and check if it is exact. If it is, terminate; the optimal solution has been found. If not, go to Step 7.

Step 7.

The least cost problem chosen from the master list in Step 6 has a facility which was chosen to be branched upon in Step 5. Set up problems $a_1 \ldots a_{Bi}$ and a_0, where B_i is the number of straight lines used to approximate the concave function of the facility.

Problems $a_1 \ldots a_{Bi}$ would have the facility location i, the corresponding pseudofacility with a fixed charge and a linear processing cost introduced into the set T; the corresponding y_j is set to one.

For problem a_0, all the pseudo facilities at the location i will have their y_j's set to zero.

Repeat Steps 3–5 and then go to Step 6.

3.5 DEMONSTRATION MODEL I

This model is the same as the general model described earlier except that the capacities of all the facilities are infinite. This situation arises in the regions where there are no processing facilities and all the landfills have sufficient capacity to accommodate all the wastes generated in the planning period. If there are no processing facilities, then the facilities may be built to any size and hence, the capacity of the facilities can be considered to be infinite for algorithmic purposes. Even if such is not the case, study of such a situation may be interesting to a planner.

The subproblem to be solved at each node of the branching tree takes up a special structure. Since all the capacities in the network are infinite, the minimum cost flow problem reduces to finding the cheapest path from each source to ultimate disposal at a landfill either through an intermediate processing facility or directly, and sending all the wastes generated at the source, through this cheapest path.

Cheapest Paths Computations

Define p_j = (modified) processing cost at site j

$\quad\quad\quad$ t_{kj} = transportation cost from k to j

$\quad\quad\quad$ a_j = reduction factor at site j

LABEL1(k) = cost of cheapest path from node k to a final disposal site

LABEL2(k) = intermediate processing site that figures in the above cheap-
$\quad\quad\quad\quad\quad\quad$ est path

LABEL3(k) = landfill site used in the cheapest path. For the cheapest path
$\quad\quad\quad\quad\quad\quad$ calculations the network shown in Fig. 3.3 is used. Since the
$\quad\quad\quad\quad\quad\quad$ network is tripartite, the calculations are simpler.

Step 0.$\quad$ Initialize the labels of all nodes to $(\infty, -, -)$

Step 1.$\quad$ Label all the landfill nodes j $(p_j, -, j)$

Step 1b.$\quad$ Let j = 1

Step 2.$\quad$ Find for all $k \in ((I \cup T') \cap I) \cup S : Q_k = p_j + t_{kj} + p_k$ If
$\quad\quad\quad$ $Q_k < $ LABEL1(k) replace LABEL1(k) with Q_k and set
$\quad\quad\quad$ LABEL3(k) = j, LABEL2(k) = k

Step 2b.$\quad$ Set j = j + 1 Go to Step 2 and continue until j is greater than the
$\quad\quad\quad$ number of landfill sites.

Step 2c.$\quad$ Let i = 1 i $\in (I \cup I') \cap I$

Step 3.$\quad$ For all k $\in$ S, find $Q_k = p_i \cdot a_i + t_{ki}$. If $Q_k < $ LABEL1(k) replace
$\quad\quad\quad$ LABEL1(k) with Q_k, LABEL2(k) with i and LABEL3(k) with
$\quad\quad\quad$ LABEL3(i)

Step 3b.$\quad$ Let i = i + 1 Go to Step 3 and continue until i is greater than the
$\quad\quad\quad$ number of intermediate processing facility sites.

Step 4.$\quad$ All the sources will now have the cheapest cost and the cheapest
$\quad\quad\quad$ path to ultimate disposal. Send all the wastes generated through
$\quad\quad\quad$ the path and compute the total cost of transportation and process-
$\quad\quad\quad$ ing.

3.6 DEMONSTRATION MODEL II

In some regions, there may be some processing facilities already in existence and hence have a finite capacity. All the landfills may have sufficient capacity to accommodate all the wastes generated in the region during the planning period. Additional facilities may be constructed to any capacity desired.

When such is the case, the processed wastes from the intermediate processing facilities can all be sent through the same path to ultimate disposal. There is no capacity restriction on the path from an intermediate facility to any landfill. This special feature is used to advantage in solving the subproblems at each node of the branching tree.

The solution procedure for the subproblem is as follows. The network we will be dealing with is shown in Fig. 3.6. There are two sets of nodes S and F besides the supersource and the supersink nodes. S is the set of all sources of waste generation and F is the set of all sites of intermediate facilities and landfills.

The arc $(0, s_i)$ $s_i \in S$ has a capacity equal to the amount of waste generated at the source s_i, a lower bound of zero and zero cost. The arc (f_j, t) $f_j \in F$ has a capacity of facility f_j and a lower bound of zero. If this is an intermediate facility site and a pseudofacility is 'built,' the capacity of a pseudofacility

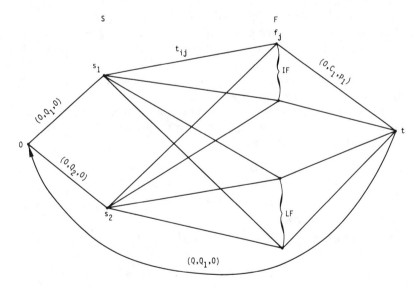

Figure 3.6.

can be taken to be equal to C_j where the straight line corresponding to this pseudofacility intersects with the next straight line of higher fixed cost as shown in Fig. 3.2. (These straight lines are the approximations to the concave cost function of a facility at the site.) Taking the capacity of the pseudofacility equal to C_j, instead of equal to the total amount of wastes generated in the region, helps increase the lower bound on the cost of the optimal solution at a node of the branching tree. If the facility site is a landfill, the cost on the arc is set equal to the (modified) processing cost at the landfill site j. This cost is obtained from the branch and bound routine. If it is an intermediate facility site, the cost on the arc (f_j t) is equal to the cost of the cheapest path from the intermediate (pseudo) facility to any landfill. The cost will include the modified processing cost at facility site j, the transportation cost from the intermediate facility site to a landfill site and the (modified) processing cost at the landfill. The cost is expressed in terms of dollars per ton of input to the intermediate facility rather than as dollars per ton of output from the intermediate facility. The reduction factor at the intermediate facility is used to compute this cost.

The arc (t 0) has a lower and upper bound equal to the total amount of wastes generated in the region and has a zero cost.

Now, the network problem is a simple transportation problem and can be solved by Fulkerson's out-of-kilter algorithm. The out-of-kilter method is used because we can start with the solution of a previous node, when solving for a particular node in the branching tree and this would save up considerable amount of computational time, because the solutions of two neighbouring nodes in the branching tree are not very different from each other.

Let us consider in detail, how the out-of-kilter algorithm can be used to advantage in the branch and bound procedure.

The solutions, primal and dual solutions, of the problems at all 'live' nodes are kept in storage. When a particular facility site is branched upon, one of the four cases can occur:

(1) The facility site is an intermediate facility site and the decision to build one of the pseudofacilities on the site is taken.

(2) The facility site is an intermediate facility site and the decision is to build no facility on the site.

(3) The site is for a landfill and the decision is to build a pseudofacility at the site.

(4) The site is a landfill site and no facility is to be built on the site.

When we branch from node A to node B, the problem to be solved at node B is only slightly different from that of node A.

Case 1. When a decision is made to build a pseudofacility at an intermediate facility site j, of the arcs of the network in Fig. 3.6, only the arc (j t) is affected. It is affected in the sense that the cost of the arc is changed. The new cost on this arc is calculated as the cost on the arc in the network corresponding to node A minus the (modified) processing cost at the facility site j at node A plus the processing cost at facility site j corresponding to the pseudofacility which is considered built for the problem of node B.

The only arc that is out-of-kilter, when the solution of node A is used as the starting solution for node B will be the arc (j t). And out-of-kilter routing is used to bring this arc back into kilter.

It should also be noted here that, the change in the fixed cost incurred at the facility site j because of the decision to build a pseudofacility must be taken into account.

Case 2. This is similar to Case 1. Here the decision is to build no facilities at the intermediate facility site j. Increase the cost on the arc (j t) to infinity (or decrease its capacity to zero) and use the out-of-kilter algorithm to bring this arc into kilter employing the solution of node A as the starting solution.

Case 3. This is a more complicated case. When a pseudofacility is built on the landfill site j, then not only is the arc (j t) going to be thrown out-of-kilter when the solution of node A is used as a starting solution, but some of the arcs corresponding to intermediate facility sites may also be. This is because when a pseudofacility is 'built' at a landfill site j the cheapest path and its cost from an intermediate facility site to ultimate disposal may change. The cheapest path calculations are carried out for all intermediate facility sites, and those intermediate facility arcs whose costs change will be out-of-kilter.

Case 4. When a landfill site is closed, the situation is similar to Case 3. The capacity on the arc corresponding to the landfill site is set at zero and its cost set at infinity. The costs on the arcs corresponding to intermediate facility sites are recomputed setting the processing cost at the landfill site as infinite. Again, the out-of-kilter algorithm is used to bring the arcs back into kilter.

3.7 DEMONSTRATION MODEL III

In many regions, there are some intermediate processing facilities in existence and hence their capacities are finite. The landfills in the region may also have finite capacities, capacities lower than the total amount of wastes generated in the region.

When we have capacity restrictions both at the intermediate facilities and the landfills, the subproblem to be solved is more complicated than that of static model I and II. The problem can be restated as a minimum cost network flow problem wherein the network flows in some of the nodes are not conserved. The solution procedure used to solve the subproblem will be a direct specialization of Jewell's algorithm (1962) to solve a more general problem of cost minimization in networks with gains.

In describing the algorithm, much of Jewell's notation and presentation will be used.

Primal Problem:

$$\min \rho = \sum_{ij} C_{ij}\, f_{ij} \tag{9}$$

subject to

$$\sum_j f_{ij} - \sum_j k_{ij}\, f_{ij} = \begin{cases} Q & i = 0 \\ 0 & i \neq 0 \end{cases} \tag{10-a}$$

$$0 \leq f_{ij} \leq M_{ij}$$
$$i,j = 0, 1, 2, \ldots, N \tag{10-b}$$

Dual Problem:

$$\max \tau = QV_0 - \sum_{ij} M_{ij}\, U_{ij} \tag{11}$$

s.t.

$$V_i - k_{ij}\, V_j - U_{ij} \leq C_{ij} \tag{12-a}$$
$$U_{ij} \leq 0 \tag{12-b}$$
$$V_i \text{ unrestricted} \tag{12-c}$$

where C_{ij} = cost on arc ij
  M_{ij} = capacity on arc ij
  Q = total amount of wastes to be disposed
  k_{ij} = reduction factor on arc ij $0 \leq k_{ij} \leq 1$

In our problem, k_{ij} takes a value between 0 and 1 if the arc corresponds to an intermediate facility and assumes a value 1 for all the other arcs. In the general problem, Jewell assumes k_{ij} are arbitrary nonzero constants, positive or negative. Another difference between the general problem Jewell considered and our problem is that the network is more 'ordered' in our problem as shown in Fig. 3.4.

The dual problem as a simple constraint (12-a) for each branch (i j), in terms of dual variables for each node, V_i and V_j, and a dual variable for each branch U_{ij}.

From the weak theorem of complementary slackness (Jewell, 1962), the following relationships exist between the primal and dual variables at optimality:

If $V_i - k_{ij} V_j < C_{ij}$, $(U_{ij} = 0)$ then $f_{ij} = 0$ (13-a)
If $0 < f_{ij} < M_{ij}$ then $V_i - k_{ij} V_j = C_{ij}$ and $U_{ij} = 0$ (13-b)
If $U_{ij} > 0$, $(V_i - k_{ij}V_j = C_{ij} + U_{ij})$, then $f_{ij} = M_{ij}$ (13-c)

We can now define three mutually exclusive 'states' of a branch as:

$$\text{A branch (i j) is} \left\{ \begin{array}{l} \text{inactive} \\ \text{active} \\ \text{hyperactive} \end{array} \right\} \text{if } V_i - k_{ij}V_{ij} \left\{ \begin{array}{l} < C_{ij} \\ = C_{ij} \\ > C_{ij} \end{array} \right\}$$

These dual states of a branch are also collectively exhaustive since dual feasibility is maintained throughout the algorithm. The optimality diagrams for each kind of the arc of a network in Fig. 3.4, are shown in Fig. 3.7. It should be noted that the arc (i' i'') is always active since $V_{i'}$, $-k_{i'i''}V_{i''}$ is always maintained equal to zero in the algorithm.

Algorithm:

1. Select an initial feasible solution to the dual of the optimal flow problem.
2. Using the principle of complementary slackness, define a restricted primal problem, and identify the primal variables which can or cannot be changed in the next step.
3. Maximize flow into restricted primal network by using the maximal flow subroutine, thus solving the restricted primal and its dual.
4. If the flow input equals Q, the optimal solution has been reached. Terminate the algorithm.
5. Otherwise, define a new feasible dual using restricted dual variables. Go to Step 2.
6. If no changes can be made in the dual, and a new feasible dual cannot be defined, the optimal flow problem is infeasible. Stop.

Step 1.
A starting feasible dual solution can be selected as follows: Referring to Fig. 3.4,

a. Set $V_t = 0$
b. for all $l \in L$ set $V_1 = C_{lt}$
c. for all $i'' \in I''$ set $V_{i''} = \min_{l \in L} (V_1 + C_{i''l})$

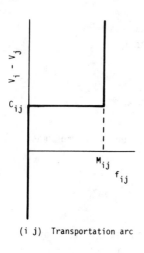

(i j) Transportation arc

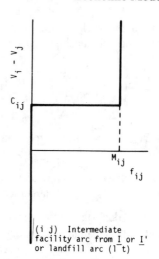

(i j) Intermediate
facility arc from $\underline{I}$ or I'
or landfill arc $(\overline{1}\,t)$

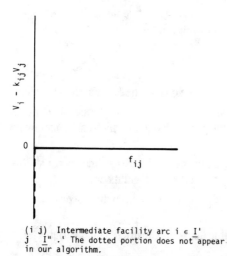

(i j) Intermediate facility arc $i \in \underline{I}'$
j $\underline{I}''$.' The dotted portion does not appear
in our algorithm.

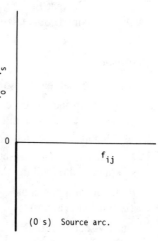

(0 s) Source arc.

Figure 3.7. Optimality diagrams.

d. for all $i' \in I'$ set $V_{i'} = k_{i'i''}V_{i''}$

e. for all $i \in I$ set $V_i = V_{i'} + C_{ii'}$

f. for all $s \in S$ set $v_\sigma = \min_{j \in I \cup L} (V_s + C_{sj})$

g. set $V_0 = \min_{s \in S} V_\sigma$

h. set all $U_{ij} = 0$.

(It should be noted here that C_{ij} in this algorithm refers to t_{ij} or P_j, defined previously, depending on whether the arc is a transportation arc or a facility arc. k_{ij} has replaced a_j for facility arcs.)

Step 2.
From Step 1, or from the output of the previous cycle (Step 5), define the following mutually exclusive and collectively exhaustive states for each branch.

A branch is hyperactive if $U_{ij} > 0$ $\hspace{4cm}$ (14-a)
A branch is active if $V_i - k_{ij} V_j - U_{ij} = C_{ij}$ and $U_{ij} = 0$ $\hspace{1cm}$ (14-b)
A branch is inactive if $V_i - k_{ij} V_j - U_{ij} < C_{ij}$ and $U_{ij} = 0$ $\hspace{1cm}$ (14-c)

The restricted primal problem is:

$$\text{maximize } F_0 \tag{15}$$

s.t. $\hspace{2cm}$
$$\sum_j (f_{ij} - k_{ij} f_{ij}) = \begin{cases} F_0 & (i = 0) & \text{(16-a)} \\ \\ 0 & (i \neq 0) & \text{(16-b)} \end{cases}$$

$$0 \leq f_{ij} \leq M_{ij} \quad \text{for all active branches} \tag{16c}$$

$$0 = f_{ij} \qquad\qquad \text{for all inactive branches} \tag{16d}$$

$$f_{ij} = M_{ij} \qquad\qquad \text{for all hyperactive branches} \tag{16e}$$

The dual restricted problem is:

$$\text{minimize} \sum_{i,j} M_{ij}\, \sigma_{ij} \tag{17}$$

s.t. $\hspace{1cm}$ $\sigma_i - k_{ij}\, \sigma_j - \sigma_{ij} \leq 0$ for all active or hyperactive $\hspace{1cm}$ (18-a)
$\hspace{4.5cm}$ branches

$$\sigma_i = \begin{cases} +1 & (i = 0) & \text{(18-b)} \\ \\ \text{unrestricted} & (i \neq 0) & \text{(18-c)} \end{cases}$$

$$\sigma_{ij} \geq 0 \quad \text{for all active branches} \tag{18-d}$$

Step 3.
The restricted primal and dual are solved by using maximal flow subroutine described below. The f_{ij} determined on the last cycle of the algorithm is used

as a starting solution. The output of the subroutine will be f_{ij} which satisfies (16) and complementary slackness will be maintained.

Step 4.
If F_0 attains the desired value Q in Step 3, the algorithm is terminated, with the f_{ij} just found as the optimal solution to the optimal flow problem (9) and (10). The dual feasible solution at the beginning of this cycle is the optimal solution to the dual of optimal flow problem, (11)–(12).

Step 5.
Otherwise, find new feasible dual variables V_i, and $U_{i'j}$:

$$V_{i'} = V_i + \theta\sigma_i \qquad (19\text{-}a)$$

$$U_{i'j} + \theta\sigma_{ij}, \qquad i,j = 0, 1, 2, \ldots, N \qquad (19\text{-}b)$$

where

$$\theta = \min\left\{ \min\left[\frac{a_j - (V_i - k_{ij}V_j - U_{ij})}{\sigma_i - k_{ij}\sigma_j - \sigma_{ij}} \right] ; \min\left[\frac{U_{ij}}{-\sigma_{ij}} \right] \right\} (19\text{-}c)$$

for all branches such that the denominators in (19-c) are positive; if none of the denominators are positive for any of the branches in the network, set $\theta = +\infty$ and go to Step 6. Otherwise θ is positive and finite; begin a new cycle of the algorithm using the new dual variables.

Step 6.
If $\theta = +\infty$, terminate the algorithm, since no feasible solution to the optimal flow restriction exists and the dual functional is unbounded.

The Maximal Flow Subroutine: (Used in Step 3)
1. Begin with f_{ij} of the previous cycle; set the label of the source = as LABEL1(0) = $-$; LABEL2(0) = ∞, LABEL3(0) = 1 where LABEL1(i) gives the arc by which node i is reached and labelled; LABEL2(i) is the maximum allowed flow into node i through the path, traced by labelling routing, from 0 to i; LABEL3(i) gives the cumulative effect of reduction factors on the path from 0 to i. For all the other nodes i, set LABEL 1(i) = $-$; LABEL2(i) = ∞ LABEL3(i) = 2.
2. a: Starting with node i, which has been just labelled (or returned from Step 7) with LABEL1(i) = I_jLabel2(i) = f_i and LABEL3(i) = k_i, consider all the arcs incident on this node, which have not been already examined since this node was labelled last time (the arcs which have been examined since this node was labelled last, need not be considered) for labelling. If the

arc is not active, store this arc as 'examined' and go to the next arc. If there are no more arcs to be examined, go to Step 6.

b. The arc I, being examined for labelling is an active branch. I will be a negative number if the arc is (j i) and positive if it is (i j).

If (i) (i j) is active and saturated or

if (ii) (j i) is active and empty then no labelling of node j is possible. Repeat Step 2 for some other arc incident on i.

c. Otherwise, node j has a labelling possibility. Calculate the tentative LABEL3(j) as LABEL2(i) · k_{ij} if I is positive, or as LABEL2(i)/k_{ji} if I is negative. (1). If LABEL3(j) is less than the tentative label computed above, node j cannot be labelled. Pick another arc incident on i and go to Step 2a. (2). If LABEL3(j) is equal to 2, label j as:

$$LABEL1(j) = 1$$

$$LABEL2(j) = \begin{cases} [\min \{LABEL2(i), (M_{ij} - f_{ij})\}] \; k_{ij} & \text{if } I > 0 \\ [\min (LABEL2(i), f_{ij})]/k_{ij} & \text{if } I < 0 \end{cases}$$

$$LABEL3(j) = \begin{cases} k_{ij}LABEL3(i) & \text{if } I > 0 \\ LABEL3(i)/k_{ij} & \text{if } I < 0 \end{cases}$$

3. (i) If LABEL3(j) is not equal to 2 and LABEL2(i)k_{ij} for LABEL2(i)/k_{ji} if I is negative) is less than LABEL3(j), then erase LABEL3(j). All the arcs incident on j will now be considered 'not examined.' If erasing the label with larger value of LABEL3(j) also cause the other labelling candidate to be erased (i.e., the newly found alternative label is a 'later' member of a labelling sequence from the source which has the original label as an 'earlier' member) then proceed to Step 4 of this subroutine.

(ii) If erasing the label with larger magnitude does not cause the other labelling candidate to be erased, then assign new labels to node j as given in 2.c.(2). Now, proceed to 2.a. with j as the node under consideration if j is not t, the supersink. If j is t, we have found an augmenting path. Go to Step 5.

4. An increase in flow into the network is possible through the structure detected in 3.(i) above. Fig. 3.8a, shows the absorbing network structure. The circles in the diagram refer to the nodes of the network and the arrows indicate the direction of labelling, not flow. It should be noted that the general form of an absorbing network is a looplike structure in which node F is a 'feedback point' for a one-way loop.

Let the two potential labels for node F be k_F^1 and k_F^2 (these are for LABEL3(F)).

a. Find KLOOP $= k_F^2/k_F^1$.

b. In the absorbing network, we have to find the maximum flow that can be circulated. From Fig. 3.8b, if x and z are the flows into node F and Y the flow out of node F, we have:

$$x + z = y$$

$$z = \text{KLOOP*Y}$$

$$x + z = z/\text{KLOOP}$$

$$x = z \left(\frac{1}{\text{KLOOP}} - 1 \right) \text{ (node KLOOP} < 1)$$

We already have the maximum flow possible into node F from supersource 0 through the path traced in labelling routine. Let us find the maximum flow

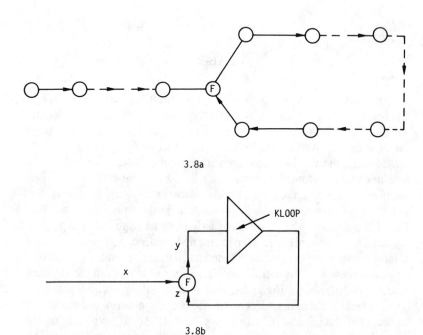

3.8a

3.8b

Figure 3.8. Absorbing structure.

that can be circulated in the loop in terms of z. Starting from node F and going around the loop in the direction of labelling, find out the maximum flow z possible.

c. Find $x^* = \min \left\{ LABEL2(F), \; z^* \left(\dfrac{1}{KLOOP} - 1 \right) \right\}$ .x^* gives the maximum flow into F through the path from the supersource. Trace back the path from F to origin and change flows as explained in Step 5 of this subroutine.

d. Compute $z^* = x^*/ \left(\dfrac{1}{KLOOP} - 1 \right)$. Starting from node F and going in the reverse direction of labelling in the loop, change flows in the loop arcs, as explained in Step 5 of this subroutine.

5. We have found an augmenting path from 0 to t. Let $i = t$, DELTA = LABEL2(i).

a. Find the arc I by which i was labelled. If I is positive, increase the flow on arc I by DELTA. The other end of I is j. Let $i = j$ DELTA = DELTA/k_{ij}. Go to 5.c.

b. If I is negative, decrease the flow on the arc by DELTA. The other end of the arc is j. Let DELTA = DELTA*k_{ij} and $k = j$ and go to 5.c.

c. If node 0 has been reached, flow augmenting is completed to 1 of this subroutine. Otherwise Go back to 5.a.

6. We have not found an absorbing structure or an augmenting $0 - t$ path and all the arcs incident on node i have been examined. Find the node from which i is labelled and call it i. If this is 0, then we have t change dual variables. If it is not 0, go the Step 2.

3.8 CALCULATIONS OF DUAL VARIABLES

At the end of max-flow procedure in Step 6, certain modes have labels, which progressed from the source to those nodes via active branches. If there are alternative labels for LABEL3(.) of a node, they either have all LABEL3(.) equal, or the node has the label affixed with the smallest value of LABEL3(.).

Step 5 of the algorithm (not of this subroutine) gives formulae for changing dual variables. It reduces to the following:

The set of nodes labelled is defined as I and the set of unlabelled nodes is defined as $\bar{I}$. Consider all the inactive and hyperactive arcs across the cut $(1,\bar{I})$ and also the inactive and hyperactive arcs connecting nodes within I. These are the only arcs which may contribute to the computation of θ in (19c). And the arcs which are considered all have k_{ij} equal to 1 in our network.

Compute:

$$0 = \min \begin{cases} \begin{aligned} &\min_{\substack{i \epsilon I \\ i \epsilon I}} \quad (k_i(C_{ij} - (V_i - V_j)))\,; \\ &(ij)\ \text{inactive} \\[1em] &\min_{\substack{j \epsilon I \\ j \epsilon I}} \quad k_j(V_i - V_j - C_{ij})\,; \\ &(ij)\ \text{hyperative} \end{aligned} \end{cases}$$

$$\min_{\substack{i,j \epsilon I \\ (ij)\ \text{inactive} \\ k_i < k_j}} \frac{C_{ij} - (V_i - V_j)}{\left(\dfrac{1}{k_i} - \dfrac{1}{k_j}\right)} \quad \min_{\substack{i,j \epsilon I \\ (ij)\ \text{hyperactive} \\ k_i > k_j}} \frac{V_i - V_j - C_{ij}}{\left(\dfrac{1}{k_j} - \dfrac{1}{k_i}\right)} \Bigg\}$$

Once we find θ, for all nodes in I let $V_{i'} = V_i + \theta/k_i$ and let all other nodes retain their node numbers. We are now ready to Step 1 of a maximal flow subroutine.

The proof of this algorithm is given in Jewell (1962). The changes that are made in the labelling routine and in the calculation of dual variables to suit the special network we have can be found true.

While the development of an out-of-kilter routine for this problem appears to be potentially useful, especially in a branch and bound procedure as we have, it is not without disadvantages. To make use of the out-of-kilter method, we have to retain the solutions of 'live' nodes of the branching tree and this may be prohibitively space consuming. Secondly, the out-of-kilter routine may be more difficult to program, especially since the primal dual method given above is itself quite complicated for programming, and hence the efficiency of computation may not be what one would hope for, with the additional memory requirements. Using a regular linear program in place of this complex network algorithm may be computationally superior when space can be found to store the solutions of live nodes. In view of these doubts, an out-of-kilter routine was not developed for this problem.

3.9 DATA REQUIREMENTS FOR CASE STUDIES

The elaboration of the case studies to follow requires access to and acquisition of data sets, such as

(1) Locations of landfills, incinerators, transfer stations and other Intermediate Processing Facilities (IPFs).

(2) Maximum capacities of the facilities to be considered.

(3) Existing and proposed transportation routes from each waste source to each disposal site, and from each IPF to a landfill site.

(4) Average round trip travel from each route for an average vehicle from each waste generation source.

(5) Cost per ton/hr for shipping waste in collection vehicles, currently and projected.

(6) Facility processing costs (capital, operating, maintenance cost) or estimates thereof for IPFs (including new waste treatment technologies).

(7) Population and per capita regional waste production and prediction for the planning period.

(8) Capital, operating and maintenance costs of landfill sites.

In order to respond to 'what-if' questions we can use a variation of data sets to do extensive sensitivity analysis on the basis of our model for the various options we contemplate. On the basis of the optimization models for regional solid waste management, as applied to cases of the Boston, Munich and Nürnberg/Fürth Metropolitan Regions, we resolve some of the main problems:

• Determination of the extent of the waste disposal problem facing various metropolitan areas in the U.S. and Europe, and inferences drawn from extensive case studies of three areas

• Determination of the waste treatment technologies, and the efficiencies thereof, now employed or employable in these areas

• Determination of the relative efficiencies and regional impacts of waste management operations in any of these areas on a comparative basis.

Furthermore, we address pertinent questions, such as:

• What are the comparative advantages of waste treatment technologies with respect to direct and indirect (social damage) cost assessment?

• Is there an urban waste disposal crisis over the next twenty years in any of the metropolitan regions? If so, which categories (health hazard, property damage, etc.) are likely to be affected foremost and where?

• What is the immediate financial potential for coordinated approaches to waste management in the regional context?

- How much variation is there in the efficiency/cost effectiveness of waste management, and hence, how much scope for improvement exists through improved methods, technologies and cooperation?
- What appears to be the potential for mechanical separation and waste recycling (as opposed to voluntary pre-separation)?

In addition to focussing on specific problem areas, a major goal is to work out various options that would represent different policies for comparison between metropolitan regions. The first case, the base case, would consist of the existing situation, another case would be expanding the number of sanitary landfills, or increasing the capacity of existing landfills, accommodating for different forms of regionalization. Option 3, in addition, would consider a number of energy, resource and materials recovery technologies. A fourth option would actually consider a decrease in the use of landfills together with a competitive mix of substitute waste treatment technologies.

CHAPTER 4

Case Studies

In this chapter some case studies are provided for the static models developed in Chapter 3, using special purpose algorithms. The data used and the solutions obtained are tabulated.

4.1 MODEL I

Several areas adjacent to the city of Munich (Landeshauptstadt München) are forming the *Munich Metropolitan Region* (Map 1), consisting of six counties: Dachau, Freising, Ebersberg, München Stadt (Munich City), Starnberg and Fürstenfeldbruck. They are chosen as the study region for Model I. In general, the regions chosen are roughly equivalent to the identification of 'agglomeration centers' (Verdichtungsräume) according to regional planning guidelines by the Ministry for the Environment of the State of Bavaria. More than thirty towns or cities are situated in the Munich Metropolitan Region with a combined population of 1.5 million in the year of 1980. The data on waste generation, processing and fixed cost of various facilities are based on published materials about solid waste management, (1973)–(1983), mainly by the Bavarian Environmental Protection Office (Bayerisches Landesamt für Umweltschutz), subordinated to the Bavarian Ministry for the Environment. In some cases the data had to be recomputed and adjusted to modelling needs for our purposes. Transportation costs are computed on the basis of actual charges to the user which according to the charge profile (Gebührenspiegel) vary considerably depending on container size, transportation distance and not the least, on whether public or private transportation is available.

The models involve regional waste management options for the Munich and Nürnberg/Fürth metropolitan areas in the State of Bavaria F.R.G. The data and the solutions are given, and interesting environmental policy conclusions are obtained.

Six types of wastes have been identified and they are: (1) municipal refuse, W1; (2) industrial waste, W2; (3) prunings W3; (4) farm refuse, W4;

55

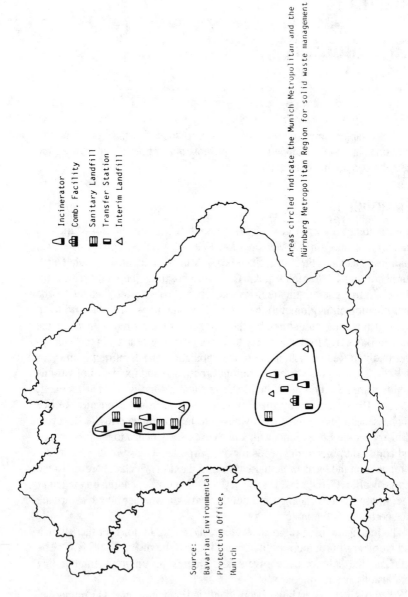

Incinerator

Comb. Facility

Sanitary Landfill

Transfer Station

Interim Landfill

Areas circled indicate the Munich Metropolitan and the
Nürnberg Metropolitan Region for solid waste management

Source:
Bavarian Environmental
Protection Office,
Munich

Map 1. State of Bavaria: Solid waste treatment and disposal facilities; December 31, 1982.

56

(5) wastes from canning and preserving industry, W5; and (6) demolition and construction debris, W6. The wastes are assumed to be generated at point sources over the region. Each point source generates only a single waste type and many point sources may be at the same geographical location.

The potential locations of intermediate facilities and landfills are chosen. There are six landfill sites, five incinerators and one combination facility (anaerobic digestion and incineration for residual waste). Table 4.1 gives the locations of these facilities and the numbers assigned to them.

Only the garbage from canning and preserving industry, waste type 5, can be treated by an anaerobic digestion plant and demolition and construction debris, W6, cannot be incinerated. The other waste types may be incinerated or sent directly to a landfill.

For each waste type, a network is formed with facility locations and sources as nodes with arcs connecting the nodes forming transportation arcs. Processing costs are associated with the nodes corresponding to the facilities.

Table 4.2 gives the waste generation data and Table 4.3 gives the weight reduction factors at intermediate facilities for each waste type. For example, a reduction factor of 90% for municipal refuse when it is sent through an incinerator plant means that if 100 tons of municipal refuse are incinerated, 10 tons of ashes will be the output.

Processing Costs

Landfills. Costs of disposing wastes at landfills are dependent on the type of waste disposed. From the cost estimations it was found that the total costs at

Table 4.1. Numbering of Facilities

Facility Type	Location	Facility Number
Landfill	Bockhorn/Erding	1
Landfill	Großlappen	2
Landfill	Dachau	3
Landfill	Landsberg/L	4
Landfill	Ebersberg	5
Landfill	Gallenbach	6
Incinerator	Munchen-Nord	7
Incinerator	Munchen-Sud	8
Incinerator	Neufahrn/Freising	9
Incinerator* (Pyrolysis)	Erding	10
Incinerator*	Weilheim/Schongau	11
Combination Facility	Geiselbullach	12

*In projection: planning horizon 1990.

Table 4.2. Waste Generation Data (tons/day)

Source No.	Quantity	Source No.	Quantity	Source No.	Quantity
W1: 1	688	W3: 21	50	39	358
2	1465	22	76	40	356
3	769	23	120	41	419
4	383	24	140	42	527
5	360	25	140	43	126
6	188				
7	150	W4: 26	7		
8	302	27	96		
9	156	28	168		
10	281	29	6		
11	355	30	355		
12	188	31	355		
		32	355		
W2: 13	158				
14	586	W5: 33	89		
15	271				
16	368	W6: 34	294		
17	16	35	828		
18	109	36	360		
19	78	37	458		
20	227	38	468		

a landfill can be conveniently divided into two parts: (1) a linear cost depending on the type of waste and (2) a nonlinear concave cost independent of the type of waste. The linear costs are lumped together with appropriate transportation costs on the arcs of the network for each waste type. The nonlinear cost function is plotted in Fig. 4.1, and is approximated by a straight line as shown. For the purposes of this example, we have a fixed cost of DM 100,000 and a variable processing cost of DM 80 for one ton/day operation for a year for all landfills. It should be noted that although landfills at different locations have different costs, the costs are reduced to a common value and any deviation from this value is lumped together with appropriate transportation costs.

Table 4.3. Reduction Factors at Facilities

Waste Type	Facility Type	Reduction %
Municipal Refuse	Incinerator	90
Industrial Wastes	Incinerator	80
Prunings	Incinerator	95
Farm Refuse	Incinerator	95
Refuse from Canning and Preserving Industry	Anaerobic Digestion	95
Demolition and Construction Debris		50

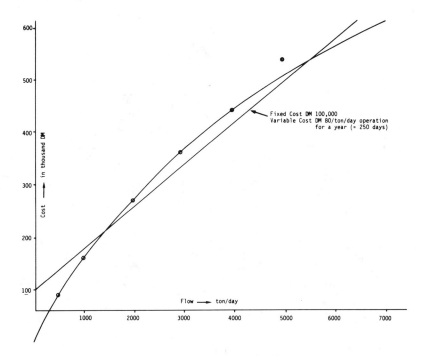

Figure 4.1. Landfill operation: nonlinear part of the cost function.

Incinerators. As in landfills, the costs at incinerators is a function for the type of waste processed. Here again, from engineering cost estimations, the total costs are classified into two parts: (1) a linear cost depending on the type of waste, and (2) a nonlinear concave cost function. Steam is generated at these incinerators and the amount of steam generated per ton of input is a function of the type of waste. Reclamation profits are also a function of the type of waste. The linear cost mentioned above, profits from steam and reclamation are all lumped together with appropriate transportation costs on the arcs of a network for each waste type. The nonlinear cost function is shown in Fig. 4.2, which also gives the approximation used in this model.

Anaerobic Digestion Plant. The cost function and its approximation are shown in Fig. 4.3.

Transportation costs, modified as explained above, are listed in Table 4.4.

All the steam produced by incineration activity is assumed to be saleable. On the basis of engineering evaluation of solid waste technology (Midwest Research Institute, 1973), the price of steam is taken to be DM 1,50/ton and

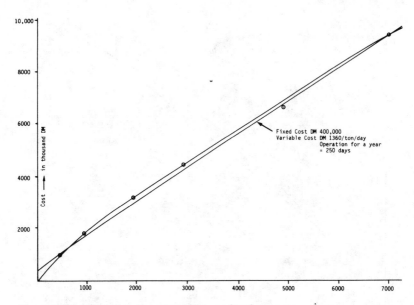

Figure 4.2. Incinerators: nonlinear part of the cost function.

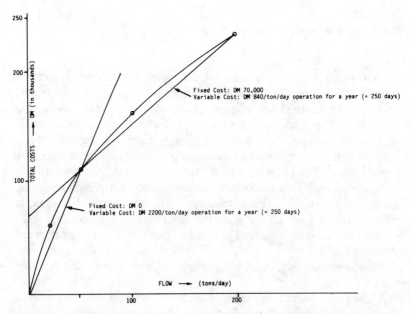

Figure 4.3. Anaerobic digestion plant as part of a combination facility: nonlinear part of the cost function.

Table 4.4. Transportation Costs[1,2,3,4] (DM/250 tons)

Dest. Origin	F1	F2	F3	F4	F5	F6	F7	F8	F9	F10	F11	F12
S1	218	375	383	154	280	281	−2135	−2014	−1913	−2176	−2005	—
S2	157	369	375	183	281	283	−2158	−2068	−1920	−2140	−2023	—
S3	131	280	368	223	289	372	−2205	−2109	−2008	−2109	−1914	—
S4	83	223	278	280	372	378	−2176	−2160	−2075	−2068	−1908	—
S5	115	221	276	284	377	383	−2141	−2176	−2068	−2015	−1742	—
S6	184	131	278	381	554	554	−2070	−2141	−2107	−1919	−1731	—
S7	277	223	130	554	378	369	−2010	−2014	−2208	−1907	−1920	—
S8	226	383	552	104	226	370	−2073	−2007	−1905	−2208	−1919	—
S9	278	549	550	130	221	254	−2108	−1921	−1907	−2176	−1923	—
S10	280	374	283	280	180	183	−2067	−1920	−2013	−2068	−2071	—
S11	369	383	372	224	131	180	−2011	−1911	−1921	−2011	−2138	—
S12	549	562	377	369	217	102	−1911	−1731	−1919	−1913	−2208	—
S13	218	375	383	154	280	281	−1894	−1773	−1672	−1935	−1764	—
S14	157	369	375	183	284	283	−1918	−1827	−1679	−1899	−1682	—
S15	115	277	281	277	368	374	−1948	−1897	−1773	−1830	−1672	—
S16	151	156	280	371	383	550	−1894	−1980	1774	−1769	−1496	—
S17	220	183	156	384	369	374	−1826	−1830	−1935	−1675	−1678	—
S18	226	383	552	104	226	369	−1832	−1766	−1664	−1967	−1678	—
S19	283	377	286	283	154	180	−1774	−1676	−1769	−1824	−1886	—
S20	549	562	377	369	217	102	−1670	−1491	−1678	−1672	−1967	—
S21	157	369	375	183	281	283	−1422	−1322	−1184	−1404	−1187	0
S22	184	131	278	381	554	554	−1335	−1405	−1371	−1183	−995	0
S23	277	223	130	554	378	369	−1274	−1278	−1472	−1171	−1181	0
S24	384	559	381	286	182	131	−1178	−999	−1178	−1180	−1463	0
S25	384	378	280	549	283	156	−1003	−1171	−1279	−998	−1375	0
S26	157	369	375	183	281	283	−913	−822	−674	−894	−677	—
S27	184	131	278	381	554	554	−824	−895	−861	−673	−481	—

Table 4.4. Transportation Costs[1,2,3,4] (DM/250 tons) (continued)

Dest.\Origin	F1	F2	F3	F4	F5	F6	F7	F8	F9	F10	F11	F12
S28	220	183	156	384	369	374	-820	-825	-930	-670	-667	0
S29	283	377	286	283	154	180	-769	-671	-764	-829	-860	—
S30	384	559	381	286	182	131	-668	-489	-668	-670	-958	—
S31	384	378	280	549	283	156	-495	-660	-769	-448	-865	—
S32	386	380	220	556	371	218	-493	-659	-819	-480	-825	—
S33	131	280	368	223	284	372	-1192	-1093	-992	-1093	-898	0
S34	218	375	383	154	250	284	—	—	—	—	—	—
S35	157	369	375	183	284	283	—	—	—	—	—	—
S36	131	280	368	223	284	372	—	—	—	—	—	—
S37	115	221	276	284	377	383	—	—	—	—	—	—
S38	184	131	278	384	554	554	—	—	—	—	—	—
S39	220	183	156	384	369	374	—	—	—	—	—	—
S40	226	383	552	104	226	369	—	—	—	—	—	—
S41	280	374	283	280	180	183	—	—	—	—	—	—
S42	369	383	372	224	131	180	—	—	—	—	—	—
S43	384	559	384	286	182	131	—	—	—	—	—	—
F7	104	278	281	224	182	374	—	—	—	—	—	—
F8	150	182	283	572	286	552	—	—	—	—	—	—
F9	276	220	150	554	384	369	—	—	—	—	—	—
F10	223	380	549	130	277	372	—	—	—	—	—	—
F11	550	563	378	371	184	115	—	—	—	—	—	—

[1] F1–F12 refer to facilities. Refer to Table 4.1 for details.
[2] Municipal wastes are generated at sources 1–12; industrial wastes at sources 13–20; pruning at sources 21–25; refuse from canning and preserving industry at source 33; and demolition and construction debris at sources 34–43.
[3] A no-entry in (ij) means transportation from origin i to destination j is not allowed.
[4] The "transportation" costs include actual transportation costs plus a linear part of processing costs at appropriate facilities. The costs are expressed in DM per one ton/day for one year operation. In other words, they are in DM per 250 tons.

Case Studies

63

Table 4.5. Optimal solution for Problem 1
FACILITY LOADINGS

Facility Type	Location	Waste Type	Quantity (tons/day)
Incinerator	Mü.-N/Mü.-S.	Municipal Refuse	5285
	Mü.-N/Mü.-S.	Industrial Wastes	1813
	Neufahrn/Erding	Prunings	517
Comb. Facility	Weilheim Geiselbullach	Refuse from Canning and Preserving Industry	89
	Total		7704
Landfill	Bockhorn	Farm Refuse	271
	Dachau/Bockh.	Demolition Debris	3122
	Dachau/Landsb.	Residue from incinerator	911
	Total		4304
Landfill	Gallenbach	Farm Refuse	1071
	Großlappen/Ebersberg	Demolition Debris	1072
	Total		2143

the waste types we have considered produce in the range of 2.5 to 3.5 tons of steam for 1 ton of input.

In the optimal solution, which is tabulated in Tables 4.5 and 4.6, waste allocation is performed through all intermediate facilities and landfills. Municipal refuse, industrial wastes, pruning and refuse from canning and preserving industry from all sources go through the incinerators (as available and projected) to the landfills associated with the incinerators. Farm refuse from some sources is partially treated by the combination facility or directly

Table 4.6. Optimal solution for Problem 1
FLOW ASSIGNMENTS

Waste Type	Sources	Intermediate Facility	Location of Int. Fac. Used	Landfill Used
Municipal Refuse	1–12	Incinerator	Mü.-N/Mü.-S.	Dachau/Gr.
Industrial Wastes	13–20	Incinerator	Mü.-N/Mü.-S.	Großl./Bock.
Prunings	21–25	Incinerator	Neufahrn/Erding	Dachau/Bo.
Farm Refuse	26–32	Comb. Fac.	—	Bockhorn/ Gallenbach
Refuse from Canning and Preserving Industry	33	Incinerator	Weilheim	Landsberg
Demolition	34–40	—	—	Dachau/La.
Debris	41–43	—	—	Großlappen/ Ebersberg

Total profits in one year operation DM 3,675,198.40.
Profit per ton of waste processed DM 0,77.

sent to the landfills. Demolition debris from some sources are sent untreated directly either to the Großlappen or the Ebersberg landfill.

With the assumption that all the steam produced could be sold, each ton of waste processed brings in a profit of DM 0.77.

The problem required 99 iterations and was solved on CDC 6600 in 7.85 seconds using Algorithm I. The time includes the time for compiling the FORTRAN program, loading, execution and printing the output.

Map 2. Waste disposal facilities in the Nürnberg metropolitan area.

4.2 MODEL II

The problem solved as an example for this model applies to an agglomeration center in Northern Bavaria, comprising the industrial region of Central Frankonia. The *Nürnberg Metropolitan Region* (Map 2) contains the city of Nürnberg, and the adjacent cities of Fürth and Erlangen in addition to more than ten smaller townships or villages. Population for this region was about 1 million in 1980. The data base is about the same as in Model I, however, unlike the previous case, because of data limitations, only one type of waste is considered.

There are twenty point sources of waste generation and Table 4.7 gives the amounts of wastes generated at these sources, in 1980. Transfer stations, treatment plants and landfills have a fixed cost of acquisition and maintenance and a processing cost proportional to the amount of wastes handled at the facility. The fixed costs, processing costs and capacities are given in Table 4.8, and Table 4.9 gives the transportation costs from the sources to the facilities to disposal sites.

An optimal solution to the problem is listed in Tables 4.10 and 4.11. Of the six intermediate facilities only the transfer station at Erlangen should be used, for additional waste generated, given by Table 4.7. Of the six disposal sites, the sites at Deponie Atzenhof, Georgensgmünd and Neunkirchen a.S.

Table 4.7. Waste generation data—Problem 2

Source No.	Quantity (tons/wk)
1	372
2	339
3	341
4	495
5	350
6	378
7	367
8	704
9	1523
10	921
11	23
12	46
13	658
14	290
15	84
16	310
17	117
18	28
19	32
20	513
21	91

Table 4.8. Cost data on facilities—Problem 2 [1,2]

Facility Type	Location	Facility No.	Fixed Cost DM/wk	Variable Cost DM/ton gener.	Capacity (tons/wk)	Reduction %
Transfer Sta.	Erlangen	1	2160	.67	5500	0
Transfer Sta.	Schwabach	2	2160	.67	5500	0
Transfer Sta.	Furth*	3	2160	.67	5500	0
Incinerator	Nurnberg-Stadt	4	12100	3.22	5000	60
Incinerator	Furth-Zirndorf	5	12100	3.22	5000	60
Incinerator	Bamberg	6	12100	3.22	5000	60
Landfill	Dep. Atzenhof	8	13000	1.04	8000	—
Landfill	Neunkirchen a.S.	9	0	4.10	14000	—
Landfill	Deponie Pyras	10	410	3.50	116	—
Landfill	Deponie Neuses	11	570	2.50	410	—
Landfill	Deponie Gosberg	12	410	3.50	116	—
Incinerator	Erlangen*	7	420	3.50	137	—
Landfill	Dep. Georgensgem.**	13	610	2.50	480	—

*In projection: planning horizon 1990.
**Expansion of present facilities projected for 1990.
[1] Costs are expressed in DM per ton.
[2] F1–F3 refer to transfer stations, F4–F7 refer to treatment plants and F8–F13 refer to centralized sanitary landfills.

Table 4.9. Transportation costs—Problem 2 (DM/ton)

Dest. / Origin	F1	F2	F3	F4	F5	F6	F7	F8	F9	F10	F11	F12	F13
S1	2.28	16.21	11.60	3.28	16.21	11.60	16.00	20.95	16.29	—	—	10.76	7.91
S2	5.44	9.62	6.57	5.44	9.62	6.57	7.23	14.53	19.47	—	—	9.00	10.84
S3	.69	7.23	6.40	.69	7.23	6.40	7.06	14.35	21.86	—	—	8.82	6.09
S4	1.54	6.11	6.30	1.54	6.11	6.30	6.96	14.25	22.98	—	—	8.72	3.85
S5	1.50	6.22	7.44	1.50	6.22	7.44	7.11	15.40	22.87	—	—	9.87	4.97
S6	3.57	5.57	8.48	3.57	5.57	8.48	5.15	16.44	23.61	—	—	10.91	7.04
S7	4.49	6.04	8.95	4.49	6.04	8.95	6.61	16.90	24.22	—	—	11.37	7.97
S8	4.28	8.45	3.87	4.28	8.45	3.87	5.11	10.34	15.24	—	—	5.84	5.05
S9	7.82	12.42	4.42	7.82	13.42	4.42	—	5.21	10.26	8.53	6.79	6.84	—
S10	5.83	6.95	9.32	5.83	6.95	9.32	5.56	15.79	20.85	—	—	11.29	5.29
S11	—	9.46	11.45	6.94	—	3.07	—	3.75	1.51	.95	17.06	—	—
S12	—	8.05	11.45	—	8.05	9.16	—	4.26	3.47	2.37	—	—	—
S13	5.84	1.97	9.24	5.84	1.97	9.24	6.63	13.82	23.93	—	—	6.79	3.63
S14	7.05	4.87	3.63	7.05	4.87	3.63	6.04	6.88	14.05	—	1.90	1.23	—
S15	8.45	5.37	5.68	8.45	5.37	5.68	3.23	10.26	20.38	—	—	3.23	7.03
S16	6.47	7.82	1.18	6.47	7.82	1.18	5.21	5.76	15.87	10.82	6.87	1.26	—
S17	5.67	4.26	4.73	—	5.76	4.73	5.92	19.43	4.26	—	9.32	10.42	2.29
S18	5.68	2.44	6.55	5.68	2.44	6.55	6.00	7.03	17.14	—	8.13	1.73	4.26
S19	—	13.38	—	9.40	—	6.99	—	3.41	4.53	.61	3.30	—	—
S20	9.87	4.26	13.27	9.87	4.26	13.27	10.98	17.85	27.96	—	—	10.82	7.66
S21	9.68	4.02	13.03	9.63	4.02	13.03	8.84	17.61	27.72	—	—	10.58	7.42
F1	—	—	—	—	—	—	—	.75	1.14	.82	.64	.46	.13
F2	—	—	—	—	—	—	—	.85	1.50	1.17	.92	.41	.21
F3	—	—	—	—	—	—	—	.92	.40	.60	.35	.15	.35
F4	—	—	—	—	—	—	—	.68	.45	.49	.38	.28	.08
F5	—	—	—	—	—	—	—	.89	.51	.70	.55	.24	.12
F6	—	—	—	—	—	—	—	.55	.24	.36	.21	.09	.21
F7	—	—	—	—	—	—	—	.42	.49	.34	.24	.36	.08

Table 4.10. Optimal solution for Problem 2
FACILITY LOADINGS

Type of Facility	Location	Quantity Processed (tons/wk)
Transfer Station	Erlangen	3990
Disposal Site	Deponie Atzenhof	6599
Disposal Site	Deponie Neuenkirchen	903
Disposal Site	Doponie Georgensmund	480

Cost of disposal per ton of waste = DM 8.17/wk.

Table 4.11. Optimal solution for Problem 2
FLOW ASSIGNMENT

Origin	Destination	Quantity (tons/wk)
Source 1	Transfer 1 (F1)	372
Source 2	Transfer 1 (F1)	265
Source 2	Disposal 6 (F13)	74
Source 3	Transfer 1 (F1)	341
Source 4	Transfer 1 (F1)	495
Source 5	Disposal 1 (F8)	350
Source 6	Disposal 6 (F13)	378
Source 7	Disposal 2 (F9)	367
Source 8	Transfer 1 (F1)	704
Source 9	Transfer 1 (F1)	1523
Source 10	Disposal 1 (F8)	921
Source 11	Disposal 2 (F9)	23
Source 12	Disposal 1 (F8)	46
Source 13	Disposal 1 (F8)	658
Source 14	Transfer 1 (F1)	290
Source 15	Disposal 1 (F8)	84
Source 16	Disposal 1 (F8)	310
Source 17	Disposal 1 (F8)	117
Source 18	Disposal 6 (F13)	28
Source 19	Disposal 1 (F8)	32
Source 20	Disposal 2 (F9)	513
Source 21	Disposal 1 (F8)	91
Transfer 1	Disposal 1 (F8)	3990

Table 4.12. Optimal solution for Problem 3
FACILITY LOADINGS

Type of Facility	Location	Quantity (tons/wk)
Disposal Site	Neunkirchen a. S.	290
Disposal Site	Dep. Pyras	422
Disposal Site	Dep. Neuses	3867
Disposal Site	Dep Gosberg	1430
Incinerator	Erlangen	1973

Table 4.13. Optimal solution for Problem 3
FLOW ASSIGNMENTS

Origin	Destination	Quantity (tons/wk)
Source 1	Disposal 6	372
Source 2	Disposal 6	339
Source 3	Disposal 6	341
Source 4	Disposal 4	495
Source 5	Disposal 4	350
Source 6	Disposal 5	378
Source 7	Disposal 3	367
Source 8	Disposal 4	704
Source 9	Disposal 4	1523
Source 10	Disposal 6	921
Source 11	Disposal 3	23
Source 12	Disposal	46
Source 13	Disposal	658
Source 14	Disposal	290
Source 15	Disposal	84
Source 16	Disposal	310
Source 17	Disposal	117
Source 18	Disposal	28
Source 19	Disposal	32
Source 20	Disposal	513
Source 21	Disposal	91

Cost of disposal per ton waste DM 6.67.

are chosen. The problem required thirteen iterations in the branch and bound procedure and an effective time of 12.1 seconds on a CDC 6600.

Algorithm II was used to solve this problem.

4.3 MODEL III

The problem chosen for this model is the same as the one chosen for Model II with the difference that the landfills have infinite capacity here. Tables 4.12 and 4.13 give the optimal solution to the problem.

In the optimal solution, none of the transfer stations or treatment plants is chosen. This is due to the fact that landfills are assumed to have infinite capacity but that their costs were taken to be the same as in Problem 2. The problem was chosen to test the correctness of the computer program and to test its efficiency and the data is not reliable or accurate.

Five of the seven landfill sites figure in the optimal solution, and since the landfills have unlimited capacity, all the wastes generated at a source go to the same landfill. Table 4.13 gives the flow of wastes from sources to landfills.

The problem required 52 iterations and was solved in 18.56 seconds on CDC 6600 using Algorithm III. As explained earlier, the time includes time for compiling, loading, execution and printing the output.

CHAPTER 5

Long Range Planning Model

In this chapter, a mathematical formulation of the long range planning of locations and expansion of facilities for regional management is presented. A procedure for solving the mathematical program is developed and an example problem is solved.

5.1 FORMULATION OF THE PROBLEM

The model developed here sets out to answer the following questions:

(1) Where should the facilities—incinerators, transfer stations, landfills, etc.—be located?
(2) When should the facilities be 'built'?
(3) What should be the capacities of facilities when they are 'built'?
(4) What are the capacity expansion strategies of existing facilities? In other words, when and by how much should the capacity of a facility be increased?
(5) How should the flows be routed from the sources to ultimate disposal each year of the planning horizon?

As in static models, we will assume that a finite number of potential locations for facilities will be chosen a priori. We will assume that capacities of intermediate facilities can only be in discrete quantities, like 500 tons/day, 1000 tons/day or 2000 tons/day, etc. This simplifying assumption is made to facilitate the solution procedure. It can be argued that this assumption is not restrictive but in fact, would better depict the real world situation.

Figure 5.1 gives the cost of setting up an intermediate facility plant as a function of its capacity. This cost is the cost incurred when the plant is erected anew. We can find the cost of expanding the capacity of an existing plant within a range from this figure. Define $E(x_1\ x_2)$ as the cost of

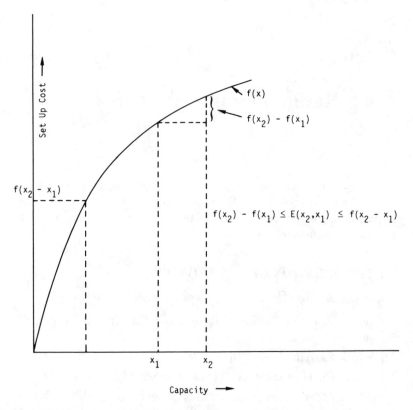

Figure 5.1. Setup cost of an intermediate facility.

expanding the capacity of a plant from x_1 to x_2, $F(x)$ as the set up cost as a function of capacity x. We have

$$F(x_2) - F(x_1) \leq E(x_1, x_2) \leq F(x_2 - x_1),$$

$F(x_2 - x_1)$ is the cost of adding a new plant of capacity,

$(x_2 - x_1)$ and clearly the cost of expansion of capacity from x_1 to x_2 cannot be greater.

It will be assumed in our model that landfills and intermediate facilities can have unlimited capacity. While the assumption that intermediate facilities can be built to any capacity is perfectly valid, the assumption on landfills may not be applicable in some regions of the country.

Having stated the assumptions behind the model, the mathematical formulation is now given.

$$\min \sum_{t \in T} \lambda_t \sum_{j \in I \cup L} \sum_{i \in S} f_{ijt} \, t_{ijt} + \sum_{t \in I'} \lambda_t \sum_{k \in L} \sum_{j \in L} f_{jkt} \, t_{jkt}$$

$$+ \sum_{t \in T} \lambda_t \sum_{i \in I \cup L} F_{it} \, (x_{it-1}, x_{it}) + \sum_{t \in T} \lambda_t \sum_{i \in I \cup L} P_{it}(x_{it}) \cdot Q_{it} \tag{1}$$

subject to:

$$\sum_{j \in I \cup L} f_{ijt} = G_{it}, \qquad i \in S \;\; \forall \, t, \tag{2}$$

$$\sum_{i \in S} f_{ijt} - a_j \sum_{k \in L} f_{jkt} = 0, \qquad j \in I \;\; \forall \, t, \tag{3}$$

$$\sum_{i \in S \cup I} f_{ikt} = Q_{kt}, \qquad k \in L \;\; \forall \, t, \tag{4}$$

$$\sum_{i \in S} f_{ikt} = Q_{kt}, \qquad k \in I \;\; \forall \, t, \tag{5}$$

$$Q_{kt} \leq x_{kt}, \qquad k \in I \cup L \;\; \forall \, t, \tag{6}$$

$$x_{kt} \in C, \tag{7}$$

$$f_{ijt} \geq 0 \tag{8}$$

where

f_{ijt} = flow from i to j in time period t
x_{it} = capacity of facility i at the beginning of time t after capacity expansion
I = set of potential sites of intermediate facilities
L = set of potential sites of landfills
S = set of sources
T = set of time periods (1, 2, ..., t)
λ_t = discount factor for time t
a_j = weight reduction factor at intermediate facility site j
$F_{it}(x_{it}, x_{it-1})$ = cost of capacity expansion from x_{it-1} to x_{it} at site i in time period t.
$P_{it}(x_{it}) \cdot Q_{it}$ = cost of processing Q_{it} units at facility i in time period t. This will depend on the capacity x_{it}.
G_{it} = quantity of wastes generated at source i in time period t.

C = set of discrete capacities which an intermediate facility can have. This
is chosen by the model user; e.g., C = (0,500,1000,1500,2000,2500)
tons/day.

The first two terms in the objective function (1) give the total discounted
transportation costs, the third term gives the total discounted capacity
expansion costs and the fourth term gives the total discounted processing
costs at the facilities.

Equation (2) implies that all wastes generated at all sources should be
shipped out. Equation (3) is a balance equation for input and output at
intermediate facilities. Equations (4) and (5) give the total quantities
processed at each facility in every time period. Inequality (6) implies that a
facility cannot process more than its capacity at any time period.

The above problem has a nonlinear objective function and a nonconvex
feasible region. This is because the third and fourth terms of objective
function (1) are nonlinear and the set C is discrete.

To solve this problem the following procedure is suggested. This
procedure does not guarantee an optimal solution to the mathematical
program (1)–(8) but will yield a solution close to optimality.

5.2 SOLUTION PROCEDURE

Step 1.

Solve each time period problem independently. Each time period problem
is that finding locations and capacities of facilities and the routing of flows,
assuming the facilities are to be 'built' from zero capacity. While solving
these problems, the restriction on capacity of an intermediate facility to be an
element of set C, is dropped. These problems fall in category of Model I of
Chapter 3 and can be solved by Algorithm I.

It is a conjecture that the first time period problem will give the routes for
the wastes for all time periods. In other words, if the waste from source i
goes through intermediate facility j to disposal site k in the first time period
problem, the same route will be used for all time periods. The conjecture was
motivated by the following facts.

(1) there are economies of scale of operation and there is no capacity
restriction on facilities; hence, it does not help to split flows from a
source into various paths.

(2) the waste generation at all sources increases with time. If it decreases, it
is conceivable that flow routing may be changed.

Step 2.

Since landfills do not exhibit economies of scale beyond a certain level of operation and since we can increase the capacity of a landfill everytime of the period without difficulty, (this would mean buying land and equipment; there are no major construction requirements) the strategy will be to increase the capacity of landfills to closely fit the demand pattern. Once the landfills are chosen from solutions of each time period problem, we will not bother about the level of activity in landfills since given the location of landfills, the cost per ton processed is almost invariant.

It is now a question of the capacity increment strategy for intermediate facilities: for obtaining a 'good' solution we will follow the heuristics that the facility locations which do not figure in the solutions of each time period problem obtained from Step 1, will not appear in the optimal solution to (1)–(8).

We have capacity requirements on intermediate facilities obtained from Step 1 but these capacities will not, in general, belong to set the C. Pick out 2 or 3 elements of the set C which are close to capacities we have for each intermediate facility for every time period. For example, if we have a capacity requirement of 740 tons/day for facility 1 in time period 1, pick 500 tons/day and 1000 tons/day as candidates for the capacity of facility 1 in time period 1.

Step 3.

At the end of Step 2, we have several possible capacities for each intermediate facility in every time period. Landfill capacities are already fixed by Step 1. The problem now is to choose the optimal capacity levels of intermediate facilities from amongst the candidates chosen in Step 2. This is done by a dynamic programming approach.

Figure 5.2 shows a three time period problem. Each node represents a possible state of the system at a particular point in time. By state of the system we mean the capacity levels of intermediate facilities and landfills.

Define $f(y,t)$ as the optimal cost of going from state y at time t to the end of planning horizon. y is a vector whose dimension is equal to the number of facilities and it is a vector of capacity levels of facilities. Define $F(0)$ as the optimal solution to the restricted problem of selecting the best combination of capacities from candidates chosen in Step 2.

$$F(0) = \min_{y} \{f(y,1) + E(C,y)\},$$

$$f(y,t) = \min_{z} \{O(y,t) + y\{E(y,z) + f(z,t + 1)\}\}$$

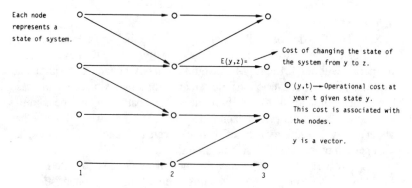

Figure 5.2. Dynamic programming approach.

where $O(y,t)$ = cost of running the system (transportation and processing costs) at time period t, given the system state y at the beginning of time period t; y = financial discount factor; z = set of possible states of the system at the time $t + 1$; $E(y,z)$ = cost of capacity expansion of the system from y to z. This is equal to the sum of capacity expansion costs of all facilities involved.

$$E(y,z) = \sum_{i \in I} E(y_i, z_i)$$

where $y = (y_1,\ldots,y_n)$, $z = (z_1,\ldots,z_n)$ and $E(y_i,z_i)$ = cost of expanding the capacity of facility i from y_i to z_i.

$O(y,t)$, the cost of operating the system in time period t with capacity vector y, can be computed by solving a transshipment problem. The problem is to minimize the total cost of processing at facilities and transportation of wastes from all sources to ultimate disposal. The facilities have finite capacity restrictions in this problem and the problem can be solved by the direct specialization of Jewell's algorithm (1962) given in Chapter 3.

The following comments on the solution procedure described above is worth mentioning:

(1) While the capacity expansion problem was restricted to intermediate facilities, it can readily be extended also to landfills.
(2) The set C can be made as large as desirable and computationally feasible. After Step 3 is carried out once, further grid optimization can be done, if necessary.

5.3 A CASE STUDY

In this section, an example problem is solved to elucidate the solution procedure described in 4.2. This time, the case study on regional management refers to the Boston metropolitan area for the period from 1970 to 2000. The case was taken from Fuertes et al. (1972). The Boston metropolitan area has thirty-nine cities and towns with a combined population over 2 million.

The planning horizon is taken to be 30 years and it is divided into three periods of 10 years each. Such a division of time is made because data on waste generation was reported for every 10 years. We could divide the planning period into any number of parts and use our solution procedure if relevant data were available. Table 5.1 gives the waste generation data. Tables 5.2 and 5.3 give the system connectivity matrix. Transportation costs are computed using a map of the Boston metropolitan area and locating sources and facilities approximately. Hence the costs considered by Fuertes et al. may be different from the costs computed here and this factor must be borne in mind when comparing the results. It is assumed that the cost of transportation by a collection truck is $267/ton/mile, the cost of transportation by a trailer truck from a transfer station is $.075/ton/mile and the cost of transportation for an incinerator reduces to $.044/ton/mile in terms of 1972

Table 5.1. Thousands of tons of residential solid waste generated/10-year period

Generating Area	1970–1980	1980–1990	1990–2000
1) Central Boston	1,050	1,210	1,340
2) Cambridge, Somerville, Charlestown	1,300	1,480	1,630
3) Malden, Revere, Chelsea Winthrop, Everett, East Boston	1,790	2,170	2,480
4) Stoneham, Wakefield, Melrose Reading	990	1,230	1,470
5) Wolburn, Medford, Winchester	770	1,010	1,170
6) Arlington, Belmont, Watertown	960	1,190	1,300
7) Bedford, Lexington, Burlington	520	710	850
8) Waltham	480	560	620
9) Newton	600	730	830
10) Brighton, Allston, Brookline	470	560	680
11) Dedham, South Western Boston	1,960	2,300	2,520
12) Milton, Quincy	770	970	1,170
13) Walpole, Norwood, Westwood	570	770	1,000
14) Braintree, Weymouth, Hingham	900	1,150	1,510
15) Canton, Stoughton, Randolph Holbrook	650	1,000	1,390
16) Wellesley, Needham	870	1,160	1,300

Table 5.2. System connectivity matrix

Generating Area	Boston Incinerator (I)	Cambridge Transfer (R)	Saugus L.F. (S)	Winchester I (F)	Watertown I (V)	Lexington L.F. (L)	Brookline I (O)	Newton I (N)	Wellesley I (W)	Dedham I (P)	Milton L.F. (G)	Quincy L.F. (Q)	Holbrook L.F. (M)	Braintree I (E)	Walpole L.F. (D)
1) Centr. Boston B	X	X	X				X	X		X	X	X			
2) Cambridge Charleston C Somerville	X	X	X	X	X		X	X							
3) Malden et al. E	X	X	X	X											
4) Stoneham et al. S		X				X									
5) Wolburn et al. R		X	X	X	X	X									
6) Arlington et al. F		X	X	X	X	X	X	X							
7) Bedford, Lexington et al. L				X	X			X							
8) Waltham T					X	X	X	X	X						
9) Newton N						X	X	X	X						
10) Brighton et al. M						X	X	X	X	X					X
11) Dedham et al. D	X						X	X	X	X	X	X			
12) Milton et al. Q	X									X	X	X	X	X	
13) Walpole et al. P										X	X	X			X
14) Braintree et al. Y										X	X	X	X		
15) Canton et al. H											X	X	X	X	
16) Needham et al.	X				X		X	X	X	X					

Table 5.3. Connectivity of transshipment network

DISPOSAL FACILITY	Braintree	Watertown	Newton	Brookline	Dedham	Winchester	Wellesley	Quincy	Milton	Holbrook	Walpole	Lexington	Saugus
PROCESSING FACILITY													
Braintree (E)								X					
Watertown (F)												X	
Boston (I)								X	X				X
Newton (N)												X	
Brookline (O)									X				
Dedham (P)									X	X			
Cambridge (R)												X	X
Winchester (V)												X	
Wellesley (W)													

prices. These figures are taken directly from Fuertes et al. (1972). The transportation costs are listed in Tables 5.4 and 5.5.

The total cost of processing and set up of landfills is shown in Fig. 5.3. The concave cost functions are approximated by straight lines and these approximations are also shown in these figures. Incinerator and transfer station costs for a 10-year period are given in Figs. 5.4 and 5.5, on the basis of technology and cost estimation in the early 1970s. Set up cost functions of incinerators and transfer stations are shown in Figs. 5.6 and 5.7, respectively.

Using the data listed above, each time period problem was solved by Algorithm I. The results are given in Tables 5.6 and 5.7.

In all the three time periods, three landfill sites at Saugus, Lexington and Milton and one incinerator plant at Newton are chosen. The routing of flows from sources to ultimate disposal is identical in all time periods and the routes are listed in Table 4.7. It should be noted here that all the wastes generated at a source are routed through the same path.

In solving the three time period problems, it was assumed that facilities are to be 'built' from zero capacity. This would mean that we incur set up costs every time period. In the case of landfills this assumption is not crucial

Table 5.4. Transportation costs

	Boston Inc.	Cambr. Trans.	Saugus LF	Winchester I	Watertown I	Lexington LF	Brookline I	Newton I	Wellesley I	Dedham I	Milton LF	Quincy LF	Holbrook LF	Braintree I	Walpole IF
Central Boston	.534	1.40	4.2				.90	2.52		2.14	1.12	1.58			
Cambridge	1.40	.267	2.80	2.11	1.20		1.34	2.11							
Malden	2.52	1.82	1.19	2.38											
Stoneham			1.54			2.66									
Woburn	1.68	1.68	2.24	.56	1.75	1.60									
Arlington	1.49	1.49	3.36	1.55	.53	1.68	2.10	1.12							
Bedford				3.10		.82		3.10							
Waltham				2.64	1.01	1.41		.83	2.54						
Newton				1.34			1.40	.40	1.87						
Brighton					.99		.35	1.12	3.26	2.54					6.14
Dedham	2.24						2.14	2.38	3.10	.347	1.34	1.95			
Milton	1.58									2.08	.53	.13	2.02	1.58	
Walpole										3.26	4.20	3.26			1.68
Braintree										2.80	2.24	2.53			5.87
Canton										3.09	2.24	3.36	.56		
Needham	3.34				2.80		2.80	1.97	.98	2.48					

80

Table 5.5. Transportation costs

	Braintree	Watertown	Newton	Brookline	Dedham	Winchester	Wellesley	Quincy	Milton	Holbrook	Wadpole	Lexington	Saugus
Braintree								.28					
Watertown												.37	
Boston								.22		.55			.55
Newton												.37	
Brookline										.74			
Dedham									.22		.46		
Cambridge												.87	.83
Winchester												.16	
Wellesley													

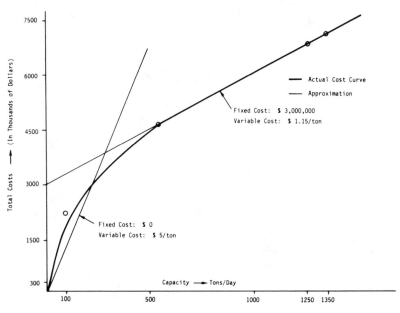

Figure 5.3. Landfill costs for 10-year period.

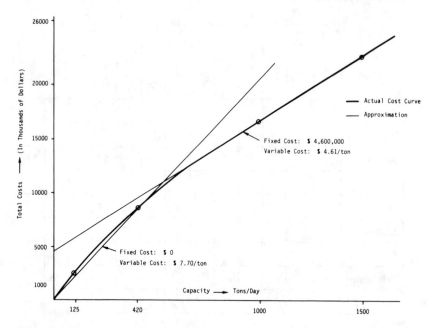

Figure 5.4. Incinerator costs for 10-year period.

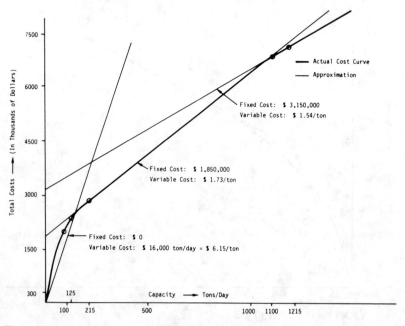

Figure 5.5. Transfer station costs for 10-year period.

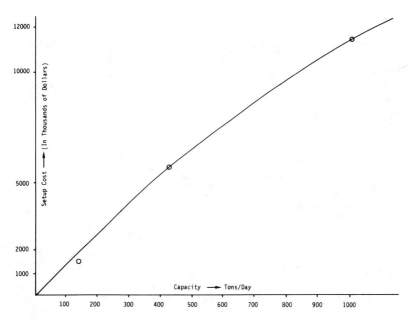

Figure 5.6. Incinerator setup costs.

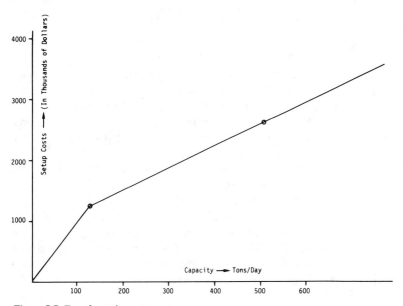

Figure 5.7. Transfer station setup costs.

Table 5.6. Facility loadings

Facility Type	Facility Location	Time Period I	Time Period II	Time Period III
Landfill	Saugus	4.080,000	4,880,000	5,580,000
Landfill	Lexington	2,924,000	3,715,000	4,221,000
Landfill	Milton	5,900,000	7,400,000	8,930,000
Incinerator	Newton	1,940,000	2,450,000	2,810,000

Total discounted cost of system operation for 30-year period = $117.78 million.

Table 5.7. Flow routes

Source	Intermediate Facility	Landfill
Central Boston	—	Milton
Cambridge etc.	—	Saugus
Malden etc.	—	Saugus
Stoneham etc.	—	Saugus
Woburn etc.	—	Lexington
Arlington etc.	—	Lexington
Bedford etc.	—	Lexington
Waltham	Newton Incinerator	Lexington
Brighton etc.	Newton Incinerator	Lexington
Dedham etc.	—	Milton
Miltin, Quincy	—	Milton
Walpole etc.	—	Milton
Braintree etc.	—	Milton
Canton etc.	—	Milton
Wellesley etc.	Newton Incinerator	Lexington

because most of the set up costs is used in buying land (part of it for buying equipment); there are no major construction requirements and the land is 'consumed.' In the case of incinerators, however, the construction and equipment would be still usable after 10 years and we should take this into account. In calculating the total discounted cost of the system for 30 years, a financial discounting rate of 4% is used.

The incinerator at Newton has a requirement of 746 tons/day, 942 tons/day and 1080 tons/day in time periods 1, 2 and 3, respectively. Assuming that the incineration will be built to a capacity of 1000 tons/day at the beginning of the planning period and will later be expanded to 1500 tons/day at the beginning of the third time period and assuming the cost of expansion from 1000 to 1500 tons/day to be at its upper bound, the total discounted cost of the system is calculated as $117.78 million. A lower bound on the optimal cost of the system can be obtained by dripping the constraints (7) and assuming that we will have 746 tons/day, 942 tons/day and 1080 tons/day capacities for the incinerator at Newton in the first, second and third time periods and by assuming that the costs of expansion will be at their lower bounds (i.e., $E(x_1, x_2) = f(x_2) - f(x_1)$). This lower bound is computed to be $115.43 million. It is clear further optimization is unnecessary, since at best we could make an improvement of $2.35 million over a 30-year period, or less than 2.3% of total expenses.

5.4 CONCLUSIONS

We modelled the multiperiod planning of regional waste management as a mathematical program (1)–(8) and suggested a heuristic procedure to solve the problem. An example involving regional management in the Boston metropolitan area for the period from 1970 to 2000 was taken from Fuertes et al. (1972) and solved by the heuristic procedure.

As conjectured earlier, we found that the first time period problem gave the routes for the wastes of all time periods. While the conjecture was found true in this one example, it cannot be taken as a fact; however, it sheds some light into the problem of multiperiod planning.

In the MIT study (Fuertes et al. (1972)), the solution proposed has a cost of $109.46 million and the cost of our solution is $117.78 million. The reason behind the difference is that the transportation costs assumed in the MIT study and here are different. The MIT study did not report the transportation costs and hence rough approximations for the costs were taken while solving the problem. It should be noted, however, that for the transportation costs and other data assumed, the solution obtained by the heuristic procedure developed in this chapter fell within $2.35 million for a 30-year period of less than 2.3% of total expenses. The computation of a lower bound on the optimal solution has been explained earlier.

CHAPTER 6

The Economics of Solid Waste Management

6.1 INTRODUCTION

Three factors have significantly contributed to a steep increase of the amount of solid waste, that make it necessary on an individual or collective basis by government to provide for the collection and disposal of these wastes. The first of these factors, population growth and its concentration in urban areas, requires proper waste removal to insure the health standards and the physical beauty of the area. The second factor, economic growth and the resulting higher incomes has meant more production, more consumption, thus more waste. Finally, technical factors, as well as changes in relative prices and tastes, have altered consumption patterns and caused increases in solid waste generation. All of these factors have had a positive effect on the demand for solid waste collection.

Where the government is involved in operating or subsidizing the collection system, demand estimation is more difficult than private sector demand estimation. In the latter case, demand theory is based on an optimization process involving indifference curves and budget constraints. It is revealed through the price system. In the public sector, preferences are only directly revealed to the extent that user charges are imposed. In cases where refuse collection is financed solely through property taxes, as in about 50 percent of American cities, the short run marginal cost to the users is zero and, hence, the true demand for the service is not revealed through various observations of price and quantity.

In the public sector, the amount of service supplied is often determined independently of market information on the demand for the service. Where the services are financed entirely through general taxation, the consumers have no way to reveal their preferences for particular services except through the political process. Their representatives must therefore act on their behalf. These decisions on allocating funds to the various services are often based solely on their beliefs of what will best serve the total community's interests. One result of this missing link between supply and demand is that some residents will be supplied more and some less service than they would be willing to pay for. In many cases, the low income areas are provided more services than they would have demanded, on the assumption that their use will benefit the high income residents who end up paying for more than the average cost of the service provided to them.

6.2 PROCEDURAL EFFICIENCY AND ORGANIZATIONAL EFFICIENCY

Efficiency can be categorized in terms of waste management: increase in efficiency in providing public service could be resulting from (1) procedural efficiency and (2) organizational efficiency e.g.:

(1) Implementing methods of operations research and management science in reducing operating costs, improving performance and providing adequate service. This would mean improving methods in vehicle routing, manpower scheduling, siting of transfer stations or disposal points. Some of the approaches using operations research models have been complicated by the fact that the problems attacked are sometimes 'wicked,' ill-structured or complex. They involve technological uncertainty, environmental constraints, awareness of the population concerned, multiple criteria problems for tradeoff valuation and related problems. They have been discussed in an extensive technical literature on engineering-economic problems in waste management: see Brill (1979), Clark and Gillian (1975), Savas (1975), Liebman (1976). Although aspects of such problems spill over in organizational areas we will not deal with such aspects here.

(2) Another aspect of increasing efficiency relates to implementing organizational changes invoking alternative information-producing and incentive-generating mechanisms for achieving better performance. A simplified dichotomy of such problem area relates to public or private provisions of services.

A large study of cities in the United States, according to Savas (1977), revealed that of those cities larger than 2,500 in population and located in metropolitan areas 37% had municipal collection, 21% had contract collection, 7% had franchise collection and 38% had private collection. According to a survey of the Bavarian Ministry for Environmental Protection in 1978, 62% of the Bavarian population is served by 111 private refuse collection firms whereas 38% are served by 58 municipal or countywise organized collection facilities. Thus in Bavaria, for solid wastes, it narrows down to either municipal collection or contract collection. Whereas contract collection is more common in smaller towns or in the countryside, municipal collection is more frequent in big cities.

I conducted a recent survey as a spinoff of waste management studies in Bavaria, which was based on a random sample of 98 Bavarian cities (population greater than 10,000). The data were obtained by questionnaires from the Bavarian Ministry for Environmental Protection, Munich. A multivariable, linear logarithmic regression model on a cross-section/time-series data set covering the period 1973 to 1980, was constructed to fit average cost per ton collected, and average annual cost per household to the following explanatory variables: amount of refuse per household generated, frequency and location of pickup, population density and hauling distance, plus a dummy indicating market structure (competitiveness) or public monopoly. Results so far are preliminary, but some highlights will be given below. The multiple correlation coefficient adjusted for degrees of freedom is 0.875, highly significant at the 0.95 probability level. Thus, by the coefficient of variation, R^2, about 76% of the variation in average annual refuse collection cost is explainable in terms of all the independent variables, of which collection frequency, pickup location and hauling distance are statistically significant.

In the light of this analysis there is reason for accepting the hypothesis that such quality variables as collection frequency and pickup location have significant positive cost effects. No significant economies of scale were revealed. The nature of the contractual arrangement proved to be a statistically significant variable. For the time period considered, and for a random sample cross section of Bavaria it was found that the costs of operating a solid waste collection system tend to be less for a private firm under contract to the community than for a municipally operated system. This finding is statistically significant at the 90% probability level. This result seems to be supported by two cases studies that have been conducted for two Bavarian cities, Munich (LH München, 1980) and Passau. In Munich we observe municipal and contract refuse collection in parallel operation though altogether municipal collection is clearly dominating. In Passau we had municipal refuse collection up to 1973 and then a switch to contract collection thereafter. In both cases we observe rather consistently over time a 10% to 15% higher average cost per ton collected for the municipal collector. However, for some unexplained reason, the gap to contract collector has been narrowing down to only 10% in the past three years.

Thus, managerial factors appear as a major cause of cost variations in solid waste collection. Government and union officials counter these findings by pointing out that productivity increase would not be the only criterion for judging performance. In a case study conducted by the German Transportation Union, ÖTV (1977), for example, a negative picture was drawn on the privatization of public services. Although the gain in productivity of private refuse collection over public refuse collection was implicitly acknowledged, however, in a biased fashion, the study points to other criteria such as wage compensation, working conditions, safety performance, work stress and specific occupational hazards as negative outcomes of privatization. On the top of it, the study argues, any productivity gains in technical innovations that have been introduced by private firms also apply to public systems (Eigenbetriebe) but that such productivity gain should result in lower prices for the public rather than in increases of profit by private refuse collectors. Finally, arguments are put forward to the effect that less efficiency is required to employ more workers. According to this line of reasoning more efficiency would lay off more workers who would be unemployed and go on welfare, so the city would be worse off. Surely, such an argument is circular because a full employment program that is based on inefficiency in government is wasteful. All government programs should be operated as efficiently as possible and if more public jobs are to be created as a matter of public policy, the additional workers should be assigned to provide additional services, such as more frequent waste collection or more street sweeping or more cleaning of parks, etc.

6.3 DEMAND FORMS

The demand for solid waste collection service by individual i in community j, d_{ij}, depends on two general factors: (1) the quantity of refuse the residents want collected; and (2) the quality of the collection service. Therefore, the unit of measurement for the level of demand is a quality-adjusted quantitative unit.

The quantity of refuse residents want collected is highly correlated with the amounts of refuse they generate. However, not all generated wastes will be collected. Some potential solid wastes will be converted to liquid wastes through the use of garbage disposal systems. Other potential collectables will be burned, perhaps because of packaging or container limits. It often requires less effort for the waste generator to burn a pile of leaves or branches than it does to bag them and place them out for collection. Also, each individual has his own propensity for littering or dumping refuse illegally.

Finally, some potentially collectable solid wastes will be allowed to decompose and will be used for soil conditioners. Each individual has his own unique relationship between the amount of refuse generated and the amount placed out for collection. These factors that influence the amount placed out for collection include income, price, and personal taste.

Each individual i in community j possesses a certain disposable income y_{ij}. It is hypothesized that for higher disposable personal income levels a higher level of service will be demanded. For instance, one would expect relatively wealthy communities to request quality-related features like backyard pickup.

Solid waste collection service is assumed to be a normal good. It is hypothesized that as the price of the service in community j, p_{ij}, rises less will be demanded. The relationship between price and quality demanded will differ from community to community, holding income constant, because different individual tastes, ta_{ij}, exist in different areas.

The final factor influencing demand is the volume of wastes generated, v_{ij}, by individual i in community j. This factor itself is a function of a number of factors discussed in the following section.

In functional form, the demand for residential solid waste collection is:

$$d_{ij} = F(y_{ij}, p_{ij}, ta_{ij})$$

where all the variables are defined above.

Due to the externalities, the total demand for solid waste collection will not be the sum of the individual demands. Individuals may feel that they could eliminate their solid waste by privately burning or dumping it, but in the aggregate, this would not be feasible due to health and safety factors. Therefore, the community demand is greater than the sum of the individual demands within the community.

6.4 WASTE GENERATION FUNCTION

The amount of waste generated by individual i in community j, v_{ij}, is an important input into the demand function. The factors hypothesized to influence the volume of solid wastes generated include climate and topography, ct_j, of community j and the disposable personal income of individual i in community j, y_{ij}.

The climate and topography factor relates to amounts of yard refuse, such as grass clippings, that will be generated. In warm, fertile areas, the per capita waste generation is hypothesized to be greater, all else equal, than in regions subject to colder temperature.

Income is also hypothesized to influence the quantity of solid waste generated. Wealthier individuals tend to purchase more items and hence, are likely to have more wastes to dispose of. Also, wealthy individuals may be less likely to conserve or recycle their wastes. In functional form the theoretical waste generation function is:

$$v_{ij} = f(ct_j, y_{ij})$$

where all the variables are defined above.

The waste generation model that was actually tested varied somewhat from the theoretical model. The theoretical model focused on what wastes were actually generated. The empirical model analyzes the wastes that will actually be placed out for collection. Although the difference in amounts may be minor, the distinction is important. The theoretical model hypothesized that volume is a function of climate and topography, and the income in a particular area. In order to test the waste generation model the volume measure used was the amount placed out for collection rather than the amount actually generated and the amount placed out for collection is not expected to be independent of the collection costs. For this reason, a price variable was also considered in the empirical waste generation model. The price variable was tested by examining those cities with user charges. Because of the lack of data on individual households, the model that was tested measured the volume for the entire city, rather than for an individual as specified in the theoretical model. The variables tested included the number of units served, income per capita, the presence of user charges, and the frequency of pickup. Some theoretical variables were not tested because they were assumed to be constant. Climate and topography are important because of the bearing they have on the amount of yard clipping and leaves.

Prior to testing the above variables, it was hypothesized that apartment units might generate less waste than single family homes because they tend to have fewer residents per unit. However, when this assumption was tested the difference between these two types of housing units did not prove to be significant. Consequently, no distinction is made in the following analysis between apartment units and single family dwellings.

Income per capita was both in the theoretical model and the empirically tested model. This variable was significant at the 95% confidence level, and indicated that in communities with higher income per capita, more refuse would be generated (and collected) per collection unit. This result conforms with the intuitive expectations that wealthier individuals purchase more and tend to purchase disposable rather than reusable products. When analyzed in the cost equations, income per capita was not a significant variable indicating that the additional wastes generated by wealthy communities are not great enough to affect costs.

6.5 SUPPLY SOURCES

The supply function for residential solid waste collection represents a combination of cost and social welfare considerations. Whereas private firms who are perfect competitors determine the amount of service supplied by the intersection of marginal cost and marginal revenue (price), public sector enterprises often consider social welfare criteria as an important determinant of supply. The lack of consideration given to marginal revenue is often due to the fact, that there may be no pricing mechanism that allows decision makers to determine the optimal levels of the various services. This lack of revenue information leads them to consider what the citizens 'need', rather than what they want or can afford. In this sense, they are forced to make a welfare judgment for the community. In the limit, all services that the community needs may not be provided because of a revenue constraint. However, the funds available for any one service are quite large when one considers the possibility of transferring funds from one budget item to another. The private supplier as well may not operate at the level where marginal cost equals price because he is likely to have a monopoly position, either by contract with the city or by law.

The social welfare judgments that are made by the city administrators are often imposed on the private suppliers as well. These considerations are based on health and safety problems involving the common good of the community and resulting from inadequate levels of service when all residents are free to choose their desired levels of service. The imposition of social welfare considerations causes the amount of service supplied to be equal to the community's "needs" which will be somewhat larger than the sum of the individual's demands.

The specific factors hypothesized to influence the supply of solid waste collection in community j, s_j, includes the inputs, x_j, the cost of the inputs, c_j (many authors prefer to combine these two factors into one), the social welfare considerations in the jth community, w_j, technology, te, including recycling considerations, the tax base, b_j, and the revenue, r_j, from any direct charges that might be imposed.

The service level, just as the level of demand is measured in quality-adjusted quantitative units. The amount of service delivered not only depends on how many tons are removed, but also on the method of removal. A higher service level can be provided by prompt, courteous collection. The amount of inputs provided is hypothesized to have a direct relationship on the service level. However, the price of those inputs is likely to bear an inverse relationship to the service level.

Each community will place a difference amount of emphasis on the social welfare aspects of solid waste collection. Some communities will place emphasis on complete and thorough solid waste collection while others will assign a relatively low priority to waste removal. The social welfare variable is intended to measure the relative priority solid waste received in the local community. The greater the weight, the more service will be provided. Similarly, the higher the level of technology available in the area of solid waste collection, the greater the level of service that can be delivered.

Finally, the level of service is hypothesized to relate directly to the tax base and the amount of revenues available from user charges. A city with a higher tax base should provide more service because resources are not as scarce in these cities. Also, a city that raises a large amount of revenue through user charges should provide higher levels of service because these revenues will likely be earmarked for the collection service.

In functional form, the supply of residential solid waste collection service is:

$$s_j = f(x_j, c_j, w_j, te, b_j, r_j)$$

where all the variables are defined above.

Although it is possible for supply to equal demand, it is not necessarily true that it must equal demand. Where the service is provided through the public sector, the decision makers may not make an accurate assessment of the community's well-being, and political pressure makes this possibility likely only in the case of urban financial crisis. Oversupply is possible to some extent in cases where the system is financed with general taxation because citizens are unaware of the costs involved for particular services. However, in the limit, citizens will demand to know where their money is spent, and administrations that oversupply services would be in as much political trouble as the undersuppliers.

Where the service is provided privately, undersupply is often controlled by the city and oversupply is controlled by the marketplace. City contracts and licenses usually stipulate minimum service levels. Providing service below this level or undersupplying the service, would violate the contract or law. Service in excess of individual demands would not sell well in the marketplace and, consequently, no incentive would exist to oversupply since no additional revenues would be likely to be granted.

6.6 PRODUCTION FUNCTION

In economic theory, the production function is a mathematical statement relating, quantitatively, the purely technological relationship between the output of a process and inputs of the factors of production, the chief purpose of which is to display the possibilities of substitution between factors of production, to achieve a given output (Shephard (1970)). The distinct kinds of goods and services which are usable in a production technology are referred to as the factors of production of that technology and, for any set of inputs of these factors, the production function is interpreted to define the maximal output realizable therefrom. Both the input and output variables are measured in time rates.

The unconstrained Cobb–Douglas production function, $y_0 = a_0 x_1 \alpha_1 x_2 \alpha_2 u_0$, is a useful expression for the solid waste collection production function. y_0 is the output measure and x_1 and x_2 are the input measures for the units of labor and capital. a_0 is a constant and u_0 is a random error term. The α_1's and α_2's which are not functions of y_0 determine the degree of scale economies. If $\alpha_1 + \alpha_2$ is less than one, equal to one, or greater than one, the production function indicates decreasing, constant, or increasing returns of scale, respectively. The Cobb–Douglas formulation has the advantage of assuming constant marginal products and is relatively straightforward in its formulation.

The Cobb–Douglas production function assumes an elasticity of substitution equal to one but there is no basis for using a production function that allows for an elasticity of substitution other than one. Based on a review of the empirical measurements of the elasticity of substitution, Nerlove (1965) concluded that there are a large number of conflicting factors operating, perhaps simultaneously, to produce the great differences observed in the results. These factors include variations in the business cycle and differences in the effective stock of capital and labor due to improvements in quality or investment of human capital.

We follow Nerlove who used the Cobb–Douglas production function in his study of the electric utilities industry with good results.

Actual empirical measurement of output in the public sector has always presented difficulties for researchers. In the particular case of solid waste, we must assume that the solid waste itself is a homogeneous output and adjust the tons collected for the quality of the collection service. The quality adjustment is necessarily arbitrary and therefore may lend itself to serious estimation errors. In short, direct estimation of the solid waste production function would be extremely difficult at best.

6.7 COST FUNCTION

Given the difficulties in working directly with production functions, a more promising strategy involves estimating the cost function. The cost function, which is derived from the reduced form of the production function, is more intuitive and leads to the identical economy of scale estimation. The only real difference one must be aware of is that estimates of economies of scale are really estimates of economies of size because, in reality, purchases of input units are not perfectly divisible.

Assuming the optimal input combinations specified in the production function are known, the cost functions express the relationship between costs and output. In general the cost function can be expressed as $c = f(y_0, p_1, p_2) + \beta$, where c is the total cost of producing y_0 units at a capital cost of p_1 and a wage rate of p_2, and β represents the costs that remain fixed in the short run. One can estimate the shape of the long run average cost curve by holding factor prices constant and regressing output against average costs for a cross-sectional set of data. From the average cost curve one can draw conclusions concerning economies of scale. For example, where the curve has a negative slope, economies of scale are present and conversely where the slope is positive, diseconomies of scale are present.

Solid waste collection involves cost other than those that rise when the service level increases. Generally, as the total payments to the inputs decrease, environmental costs increase so that, in the limit, when no one provides a collection service the environmental costs reach their peak. This differs from the typical cost function because, in solid waste collection, total (social) costs, even at zero level of output, will never go to zero. However, once the quantity and quality of the service conforms to minimum health standards, environmental costs will become negligible.

In addition to output and factor prices, average cost is also hypothesized to depend on technology, the environment within the service must be delivered, and the service level. The environmental factors include those items not under the control of the collection system operators, but likely to influence the average cost of solid waste collection. The climate and topography of an area are environmental factors likely to influence the average cost of solid waste collection. However, these factors relate primarily to the amounts of wastes generated and would only influence the average cost per unit served, rather than the average per ton collected. Population per square mile is also likely to influence the average costs. It is hypothesized that high-density areas will have a lower average cost because less nonproductive drive time will be needed. Similarly, average costs should increase as the distance to the disposal site increases.

Finally, the service level should directly influence the average cost. The higher the level of service, the higher the average cost. Therefore, if the frequency of collection increases or the collection location is shifted from the curb to the backyard, average cost is hypothesized to increase.

The theoretical cost function specified that the quantity of output, the factor prices, the degree of service, the environment within which the system must operate, and technology may influence the cost of operating a solid waste collection system. The theoretical model was tested using the specific variables: volume (which was dropped in favor of the number of units served), wage rates, pickup location, frequency of pickup, financing, type of operator (municipal or private), distance to the disposal site, crew size, and population density.

The variable "technology" was not tested because all cities were assumed to have the same level of technology. It is extremely hard to define a technological change in the solid waste collection industry. All the cities used a collection crew with a mechanical packer truck—this could be interpreted as constant technology throughout. However, some cities used packers with greater compaction ability; is this a different technology? Rather than arbitrarily deciding one or the other engineering improvements was a technological change instead of a cost minimizing response to different factor prices, all cities were assumed to be operating with the same technology. Those cities with side-loaders or high-density compaction units did not appear to have any measurable cost advantages over the other cities. This conclusion was drawn indirectly from the fact that side-loaders are operated with one and two-man crews and crewsize was not found to be a significant variable.

The variable "operator" was tested to determine whether there were significant cost differences between public (municipal) and private (contract or free competition) systems. This variable proved to be significant at the 95% level and indicated that in the aggregate, private contract payments tended to be less than the costs involved in municipally-operated systems. This does not mean that, all else equal, every private contract system would be cheaper since many municipal systems were found to be quite efficient. The typical or average city, however, could save 12.8% of its cost by switching from a municipal to a private contract system, ceteris paribus.

Prior to testing any cost variables in a regression framework, it was hypothesized that apartment units might not cost as much to collect as single family dwellings. These lower costs may result from less refuse being generated by apartment units than single family homes, as well as the use of common waste storage containers in apartment complexes. When this hypothesis was tested no statistically significant difference was observable between the cost of collecting apartment units and the cost of collecting single family homes. For this reason, no distribution is made below between apartment units and single family dwellings.

6.8 PRODUCTIVITY

Before beginning to examine some of the areas of solid waste collection where productivity analysis has been applied, a number of terms must be defined to avoid confusion.

Within micro-economics, production theory deals with the concepts of total product, average product, and marginal product. Total product is simply the measure of output as a function of the level of a particular input. When one increases an input while holding the other inputs constant, the amount of total product will usually increase first at an increasing and then decreasing rate until finally no more output can be achieved by increasing the single input.

Basic micro-economics teaches us that the average product can be derived from the total product curve or schedule. The average product of an input is the total product divided by the amount of the variable input. Thus average product is the output-input ratio for each level of output and the corresponding volume of input. The average product increases so long as the addition to the total product is greater than the previous average when the variable factor is increased.

The average product, the output-input ratio, is a measure of productivity. When the average product reaches its peak, the output per unit of input, say man-hours, is the greatest. In solid waste collection, much attention has been focussed on one particular productivity measure, output or tons collected per manhour.

The marginal product is also an important concept in the production theory. The marginal product of an input, say labor, is the addition to total product attributable to the addition of one unit of labor to the production process, the fixed input remaining unchanged. It is obvious that no collection system should ever operate in the region where an additional unit of labor would reduce output—that is where the marginal product of labor was negative. This could, in fact, happen if too many collectors were assigned to a truck and they began to get in each other's way.

Production theory requires that the nature of the product remain un-changed over the period of analysis. In a service industry, this is the same as saying that the quality of service must be held constant. Production analysis in a service industry is difficult because the unit of output is more elusive than in a process where a physical product is produced. Changing the level of service is analogous to alterations in the nature of a physical product in a manufacturing industry. For example, productivity cannot necessarily be improved by shifting part of the collection burden from the collectors to the residents. A city that discontinues backyard pickup and switches to curbside pickup will not necessarily increase the productivity of its workers by implementing this change although a cursory look at the output per manhour ratio would seem to indicate an improvement.

The concepts of cost theory must be used instead of the concepts of production theory when the input variables are examined simultaneously. In particular, the cost efficiency analysis would allow one to determine the combination of inputs that would produce the desired level of output at the least cost. This least cost point is defined to be the point of maximum cost efficiency. In the solid waste collection industry, one would find that the ratio of tons collected per monetary unit spent to collect those tons would reach its maximum for a particular level of output at the point of maximum cost efficiency. The point indicates the combination of inputs where every productive process should be striving to operate. Technically, this means the system is operating in its expansion path.

Cost inefficiencies occur when the inputs are not used in a combination that would minimize costs for a particular level of output. A city might decide to invest in some additional labor and capital to build a transfer station the objective being to reduce the operating costs of the collection system while holding the level of service constant. If the ratio of tons collected per monetary unit spent was lower after the transfer station had gone into operation, the change to the transfer station would have been cost ineffi- cient.

One additional type of analysis that is neither "productivity" analysis nor cost efficiency analysis is that type which focuses on avoiding wastage. The analysis does not revolve around examining the marginal product of labor when more or less labor is used in the collection system. Nor does the analysis focus on the appropriate combination of labor and capital. Rather, the analysis focuses on getting the most from the resources that are already present. It is necessary to address the question of wasted resources, but it must be remembered that much of that analysis is not traditional productivity analysis.

One area where productivity analysis is often applied is the analysis of crew size. This is a complex area for productivity analysis because a discussion of crew size involves both the labor and capital inputs. To provide the most useful analysis of crew size, one would want to go beyond labor productivity and look at the cost efficiency of various crew sizes.

Other, less complex factors have also been suggested for improving solid waste collection productivity. These factors include such items as routing practices, vehicle maintenance, storage containers, and management style. Unfortunately, these other factors cannot be analyzed empirically with the data available from other parts of this study.

Routing affects productivity in that good routing better utilizes the capital and labor thus producing more output with the same or less input. Routine design efficiency was certainly one of the missing variables in the cost equations, but was omitted because of the difficulty one would have in quantifying this variable.

It is important to point out that a considerable amount of work has been completed in the area of optimal routings and the locations of transfer stations. David Marks and Jon Liebman (1970) performed a comprehensive study for the U.S. Department of Health, Education and Welfare's Bureau of Solid Waste Management. Marks and Liebman successfully used a large-scale "flow of goods" model to determine the best location of transfer stations. However, after much work, they concluded that "securing an optimal solution to even a small-scale routing problem in a reasonable amount of time was difficult to impossible. Although they concluded that there is no existing optimization technique that will allow the investigation of a city-sized routing problem, they show how heuristic methods have advanced and feasible (versus optimal) solutions for quite complex problems can be found quickly. Unfortunately, no one has been able to test the difference between feasible and optimal solutions.

Vehicle maintenance and purchasing policies are also important areas affecting productivity. Having too much, too little, or non-optimally maintained capital can provide a mix with the labor input that results in less than an optimal level of productivity. When there are not enough trucks available for collection, costs are incurred paying for idle crew time or crew repair time, with no resulting output. The same problem exists when equipment breaks down on the routes. These problems can result from poor equipment, old equipment, absence of regular servicing, and poor management. Possibly, one factor that may account for the lower cost of private contractors is the vehicle maintenance and purchase policies in municipally-run systems. No hard data were gathered in this study on municipal maintenance policy, but casual observations indicate that many municipalities, caught by increasing budgetary pressures, are postponing the purchase of new equipment and tending to purchase lower quality equipment. If this is true, these short-run savings may already be resulting in higher costs in the long-run to municipal systems. The private collectors may be able to take a more realistic long-term view of their operations, especially when they have firm contracts extending into the future.

The key ingredient to high productivity is good management. Besides making decisions concerning capital purchase and utilization, the management must attempt to achieve the most output per unit of labor input. Due to the type of workers attracted to jobs in the solid waste collection area, skilled management is vital. Increasing instances of labor unrest indicate that improvements could be made to bolster worker morale and thereby, hopefully, productivity. Unfortunately, improvements in worker morale do not always lead to increases in productivity. One intervening variable is the union and the work rules they impose. It is beyond the scope of this chapter to discuss labor relations but this factor should be kept in mind when analyzing the gains that could be made in productivity by improving worker morale.

An incentive system has been suggested as a logical alternative to the eight-hour work day because of the benefits to efficiency and morale that could result. Under this system, workers would be paid for completing their routes rather than for working eight hours. If increased productivity on the routes could lead to reductions in the payroll, the reductions in man-hours could be translated into cost savings. However, there are drawbacks to this system. The service levels could drop because collectors would now have an incentive to quickly toss their cans back on the curb rather than replace them carefully. If they were paid by the hour, careful placement of the cans on the curb would be a good way to build up more time. Also, they would be less inclined to wait for the last minute residents who bring their refuse out just as the collectors approach. The second problem with this method was expressed by a private collector who switched away from the incentive system. Incentive systems have a detrimental effect on the equipment. When crews are in a hurry, they tend to accelerate rapidly and brake too quickly, which is likely to lead to maintenance problems and increased fuel usage. One author tested a payment system variable in a regression model and found it to be insignificant.

Other management practices that would be clearly beneficial to productivity include safety training programs, career ladders, and opportunities for feedback from the workers.

In summary, the optimal practices that would be clearly beneficial to productivity and cost efficiency depend on a number of factors. High productivity can result from matching the proper crew size to the correct equipment. Crew size depends, to some extent, on relative labor costs but crew size from one to four seems to make little difference in terms of cost efficiency. Scientific routing practices and proper storage containers should be used. With proper management and long-range equipment-planning, productivity can be an important contributing factor to an efficient and effective solid waste collection system.

6.9 FACILITY SITING

Economics and Political Issues

A critical issue having been technically addressed in Chapter 5 is deciding on practical waste management functions. The principal determinants are economics and political factors. Proper disposal of waste, in particular hazardous waste, is expensive and is getting even more expensive with the advent of more stringent regulatory actions.

The central siting issue with respect to economics is whether those higher costs may prohibit the ability to develop sites where and when needed. Process and costs for the various technologies vary considerably. Moreover, they are typically very expensive if compared with a benchmark such as nonhazardous solid waste disposal in sanitary landfills.

Certain general trends can be observed from the price structure of the technologies. Typically, it costs significantly more to dispose of highly toxic wastes. It also costs more to remove any ultimate liability one would incur with a given waste. That is, the technologies designed to destroy facilities has centered around the siting process itself. The lack of substantive public input during the siting process and the inadequate representation of the public on siting boards has frequently stopped siting attempts in their tracks. This perception of inadequate public representation often gave rise to a wide variety of technical and economic concerns. In some cases, concerns have been varied on a local level about the use of subjective siting criteria and the apparent availabilities of superior alternative sites. Furthermore, significant questions were raised concerning the potential economic disbenefits associated with proposed sites and the lack of compensation for these potential disbenefits.

6.10 NEED OF FURTHER ANALYSIS

One of the most important unexplored areas relating to solid waste collection is the interaction between the costs involved with collection and disposal. Cost studies similar to the current study seek to determine the significant factors affecting collection costs, but are forced by data limitations to accept a combined collection and disposal figure as their dependent variable. A methodology needs to be determined whereby collection and disposal costs would be properly separated. The variables tested in this study need to be tested against true collection cost figures. Also the important variables, both collection related and disposal related, influencing disposal costs also need to be analyzed. Ultimately, a study relating just collection costs and one focusing on disposal costs need to be merged to allow for the design of a totally integrated system of solid waste collection and disposal.

Another area needing more in depth research is the collection of user charges for a solid waste collection system. It is often difficult to determine whether the costs reported for the solid waste collection system included this cost, or whether it is part of the cost of some other city department such as the utilities. More research needs to be done on the most appropriate method of assessing and collecting these charges.

Also, a closer look needs to be given to the incentive systems used in the various collection systems. Discrepancies were noted between the number of hours the cities officially reported informally. One may suspect that many cities may now find themselves in a position where the solid waste collectors are among the highest paid hourly employees in the city when one considers the number of hours they are paid for as opposed to the number of hours they actually work. In trying to rectify this situation by rearranging routes and requiring more hours the cities may find themselves faced with serious labor problems.

Finally, additional work needs to be done on quantifying the externalities found in the solid waste collection area. It was recognized that solid waste collection service has negative externalities because the lack of collection service will injure others. These represent real costs that the cities should consider when adjusting their service levels. However, a methodology to quantify these externalities still needs to be developed.

BIBLIOGRAPHY

A. MATHEMATICAL MODELS

Altman, S.M., E.J. Beltrami, S.S. Rappaport, G.F. Schoepfle, "Nonlinear Programming Models of Crew Assignments for Household Refuse Collection," *IEEE Transactions on Systems, Man and Cybernetics*, VSMC-1, No. 3, 1971

Anderson, L.E. and A.K. Nigam, "A Mathematical Model for the Optimization of a Waste Management System," ORC 67-25, Operations Research Center, University of California, Berkeley, CA, 1967

Ayles, P.S., E.M.L. Beale et al., "Mathematical Models for the Location of Government," *Math. Progr. Study*, Vol. 9, 1978

Bellmore, M., J.C. Liebman and D.H. Marks, "An Extension of the (SWARC) Truck Assignment Problem," *Naval Research Logistics Quarterly*, Vol. 19, 1972

Beltrami, E.j. and L.D. Bodin, "Networks and Vehicle Routing for Municipal Waste Collection," *Jour. of Networks*, No. 3, 1973, 65–94

Brill, E.D., "The Use of Optimization Models in Public Sector Planning," *Management Science*, Vol. 25, No. 5, 1979, 413–422

Clark, R.M., "Economics of Solid Waste Investment Decisions," *Journal of Urban Planning and Development Division*, American Society of Civil Engr., Vol. 96, UPI, 1970

Clark, R.M. and B.P. Helms, "Decentralized Solid Waste Collection Facilities," *Journal of Sanitary Engineering Division*, ASCE, Vol. 96, No. SA5, 1970

Discussions:
(a) Heany, J.P., *Journal of Sanitary Engineering Division*, ASCE, Vol. 97, No. SA2, 1971
(b) Kirby, M., *Journal of Sanitary Engineering Division*, ASCE, Vol. 97, No. SA3, 1971

Clark, R.M. and B.P. Helms, "Fleet Selection for Solid Waste Collection Systems," *Journal of Sanitary Engineering Division*, ASCE, Vol. 97, No. SA1, 1972

Clarke, G. and J.W. Wright, "Scheduling of Vehicles from a Central Depot to a Number of Delivery Points," *Operations Research*, Vol. 12, No. 4, 1964

Coyle, R.G. and M.J.C. Martin, "The Economics of Refuse Collection," *Operational Research Quarterly* 20 (special conference issue), 1969

Dantzig, G.B. and J.H. Ramser, "The Truck Dispatching Problem," *Management Science*, Vol. 6, No. 2, 1959

Dee, N. and J.C. Liebman, "Optimal Location of Public Facilities," *Naval Research Logistics Quarterly*, Vol. 19, 1972

Esmaili, H., "Facility Selection and Haul Optimization Model," *Journal of Sanitary Engineering Division*, ASCE, Vol. 98, No. SA6, 1972

Fuertes, L.A., J.F. Hudson and D.H. Marks, "Analysis Models for Solid Waste Collection," Report 72-70, Dept. of Civil Engineering, M.I.T., Cambridge, MA, 1972

Gottinger, H.W., "A Computable Model of Solid Waste Management," *Europ. Jour. of Oper. Research,* Vol. 35, 1988.

Gottinger, H.W., "Computational Models for Regional Solid Waste Management—Part I," *Syst. Anal. Model. Simul. 3 (1986) 5,* 395–407

Gottinger, H.W., "Computational Models for Regional Solid Waste Management—Part II," *Syst. Anal. Model. Simul. 3 (1986) 5,* 507–518

Gottinger, H.W., "A Computational Model for Solid Waste Management with Applications," *Appl. Math. Modelling,* Vol. 10, Oct. 1986

Hausman, W.H. and P. Gilmour, "A Multiperiod Truck Delivery Problem," *Transportation Research,* Vol. 1, No. 4, 1967

Helms, B.P. and R.M. Clark, "Locational Models for Solid Waste Management," *Journal of Urban Planning and Development Division,* ASCE, Vol. 97, No. SA2, 1971

Helms, B.P. and R.M. Clark, "Selecting Solid Waste Disposal Facilities," *Journal of Sanitary Engineering Division,* ASCE, Vol. 97, No. SA4, 1971

Discussions:

(a) Liebmann, J.C., *Journal of Sanitary Engineering Division,* ASCE, Vol. 98, No. SA1, 1972

(b) Tanaka, M., *Journal of Sanitary Engineering Division,* ASCE, Vol. 98, No. SA1, 1972

(c) Roberts, K.J., *Journal of Sanitary Engineering Division,* ASCE, Vol. 98, No. SA2, 1972

(d) Helms, B.P. and R.M. Clark, *Journal of Sanitary Engineering Division,* ASCE, Vol. 98, No. SA6, 1972

Hung, M.S. and J.R. Brown, "An Algorithm for a Class of Loading Problems," *Naval Research Logistics Quarterly,* Vol. 25, 1978

Ignall, E.J., P. Kolesar, and W. Walker, "Linear Programming Models of Crew Assignments for Refuse Collection," *IEEE Trans.,* VSMC-2, Vol. 5, 1972

Jewell, W.S., "Optimal Flow Through Network with Gains," *Operations Research,* Vol. 10, 1962, 476–499

Liebling, T.M., "Graphentheorie in Plangungs- und Tourenproblemen am Beispiel des Städtischen Straßendienstes," *Lecture Notes in Operations Research and Math. Economics,* Springer: New York, 1970

Liebman, J.C., "Models of Solid Waste Management," Chapt. 5 in S. Gass and R. Sisson (eds.), A Guide to Models in Government Planning and Operations, Sanger Books, Potomac, Md. 1975.

Male, J.W., "A Heuristic Solution to the M-Postmen's Problem," Ph.D. Thesis, The Johns Hopkins University, Baltimore, MD, 1973

Marks, D.H., J.L. Cohen, H.L. Moore and R. Stricker, "Routing for Municipal Services," in Association for Computing Machinery, Sixth Annual Urban Symposium, NY, ACM, 1971

Marks, D.H. and J.C. Liebman, "Mathematical Analysis of Solid Waste Collections," USPHS, Bureau of Solid Waste Management, 1970

Marks, D.H. and J.C. Liebman, "Optimal Location of Solid-waste Transfer Stations," 37th National Meeting of ORSA, Wash., D.C., April 1970

Marks, D.H. and J.C. Liebman, "Locational Models: Solid Waste Collection Example," *Journal of Urban Planning and Development Division,* ASCE, Vol. 97, No. SA2, 1971

Marks, D.H.J. and R.M. Stricker, "Routing for Public Service Vehicles," *Journal of Urban Planning and Development Division,* ASCE, Vol. 97, No. SA6, 1971

Morse, N. and E.W. Roth, "Systems Analysis of Regional Solid Waste Handling," USPHS, Bureau of Solid Waste Management, 1970

Nigam, A.K., "Optimal Strategies in Capacity Expansion," ORC 70-20, Operations Research Center, University of California, Berkeley, CA, 1979

Pierce, J.F., "Direct Search Algorithms for Truck Dispatching Problems," *Transportation Research,* Vol. 3, No. 1, 1969

Ross, G.T., "A Branch and Bound Algorithm for the Generalized Assignment Problem," *Math. Programming,* Vol. 8, 1975

Rossman, L.A., "A General Model for Solid Waste Management Facility Selection," Department of Civil Engineering, University of Illinois, Urbana, IL, 1971

Sethi, S.P. and J.H. Bookbinder, "The Dynamic Transportation Problem: A Survey," *Naval Research Logistics Quarterly,* Vol. 27, 1980

Stricker, R.M., "Public Sector Vehicle Routing: The Chinese Postman Problem," Unpublished M.S. Thesis, Department of Electrical Engineering, M.I.T., Cambridge, MA, 1970

Tanaka, M., "A Linear Programming Model for Solid Waste Management," Ph.D. dissertation, Field of Civil Engineering, Northwestern University, Evanston, IL, 1970

Wahi, P.N. and T.I. Peterson, "Management Science and Gaming in Waste Management," *Journal of Sanitary Engineering Division,* ASCE, Vol. 98, No. SA5, 1972

Fulkerson, D.R., "An Out-of-Kilter Method for Minimal Cost-Flow Problems," *J. Soc. Industrial and Applied Math.,* Vol. 9, 18–27, 1961

Walker, W., Aquiline, M. and Schur, D., "Development and Use of a Fixed Charge Programming Model for Regional Solid Waste Planning," Rand Corp. Paper 5307, Santa Monica, Oct. 1974.

B. COMPUTER SIMULATION MODELS

Berthouex, P.M. and L.C. Brown, "Monte Carlo Simulation of Industrial Waste Discharges," *Journal of Sanitary Engineering Division,* ASCE, Vol. 95, No. SA5, 1969

Bodner, R.M., E.A. Cassell and P.J. Andros, "Optimal Routing of Refuse Collection Vehicles," *Journal of Sanitary Engineering Division,* ASCE, Vol. 96, No. SA4, 1970

Discussion:
(a) Marks, D.H. and J.C. Liebman, *Journal of Sanitary Engineering Division,* ASCE, Vol. 97, No. SA1, 1971
(b) Stearns, R.P., *Journal of Sanitary Engineering Division,* ASCE, Vol. 96, No. SA6, 1970

Clark, R.M. and J.I. Gillean, "Analysis of Solid Waste Management Operations in Cleveland, Ohio: A Case Study," *Interface,* No. 6, 1975, 32–42.

Clark, R.M., B.L. Grupenhoot, B.L. Garland and A.J. Klee, "Cost of Residential Solid Waste Collection," *Journal of Sanitary Engineering Division,* ASCE, Vol. 97, No. SA5, 1971

Clark, R.M., R.O. Toftner and T.N. Bendixen, "Management of Solid Waste—The Utility Concept," *Journal of Sanitary Engineering Division,* ASCE, Vol. 97, No. SA1, 1971

Esmaili, H., "Merced County Solid Waste Management Study—Phase I," report to Merced County, Association of Governments, Engineering Science Inc., Arcada, California, 1971

Esmaili, H., "Merced County Solid Waste Management Study—Phase II," 1972

Klee, A.J., "DISCUS—A Solid Waste Management Game," *IEEE Transactions on Geo-Science and Electronics,* GE-8, No. 3, 1970

Quon, J.E., A. Charnes and S.J. Wersan, "Simulation and Analysis of a Refuse Collection System," *Journal of Sanitary Engineering Division,* ASCE, Vol. 91, No. SA5, 1965

Discussion:
(a) Betz, J.M. and R.P. Steams, *Journal of Sanitary Engineering Division,* ASCE, Vol. 92, No. SA3, 1970

Quon, J.E., R.M., Mertens, M. Tanaka, "Efficiency of Refuse Collection Crews," *Journal of Sanitary Engineering Division*, ASCE, Vol. 96, No. SA2, 1970

Quon, J.E., M. Tanaka and A. Charnes, "Refuse Quantities and Frequency of Service," *Journal of Sanitary Engineering Division*, ASCE, Vol. 94, No. SA2, 1968

Quon, J.E., M. Tanaka, S.J. Wersan, "Simulation Model of Refuse Collection Policies," *Journal of Sanitary Engineering Division*, ASCE, Vol. 95, No. SA3, 1969

Rothgeb, W.L., "Computerized Refuse Collection," *Public Works,* Vol. 101, No. 4, 1970

Stroope, R.D., "Computerized Decision Analysis Under Uncertainty," M.S. Thesis, Department of Civil Engineering, University of Washington, 1970

Truitt, M.M., J.C. Liebman, C.W. Kruse, "Simulation Model of Urban Refuse Collection," *Journal of Sanitary Engineering Division*, ASCE, Vol. 95, No. SA2, 1969

Truitt, M.M., J.C. Liebman and C.W. Kruse, "Mathematical Modelling of Solid Waste Collection Policies," SU-Irq, USPH, Bureau of Solid Waste Management, 1970

C. OTHER PAPERS

DeGeare, T.V. and J.E. Ongerth, "Empirical Analysis of Commercial Solid Waste Generation," *Journal of Sanitary Engineering Division*, ASCE, Vol. 97, No. SA6, 1971

Derigs, U., Korte, B., "The branch and bound approach to combinatorial optimization problems," in M. Singh (ed.) *Encyclopedia of Systems and Control*, Pergamon Press: Oxford, New York, 1983

Gouleke, C.G. and P.H. McGauhey, "First Comprehensive Studies in Solid Waste Management," First and Second Annual Reports, USPHS (SU-3rd)), 1970

Gouleke, C.G., "Comprehensive Studies in Solid Waste Management," Third Annual Report, U.S. EPA (SU-10 rg), 1971

Gouleke, C.G. et al., "Comprehensive Studies in Solid Waste Management," Final Report, SERL 72-3, 1972

Hickman, H.L., Jr., "Planning Comprehensive Solid Waste Management Systems," *Journal of Sanitary Engineering Division*, ASCE, Vol. 94, No. SA6, 1968

B. Jäger, *"Abfallwirtschaft in der Bundesrepublik Deutschland,* (Waste use in the F.R.G.), BMFT: Bonn 1984

Johnson, St. P., *The Pollution Control Policy of the European Communities*, Graham & Trotman, London, 1983

Korte, B. (ed.), *Modern Applied Mathematics: Optimization and Operations Research*, North Holland: Amsterdam, 1982

Liebman, J.C., "Some Simple-Minded Observations on the Role of Optimization in Public Systems Decision-Making," *Interfaces,* Vol. 6, No. 5, 1976

MacFadyen, J.T., "Where will all the garbage go," *The Atlantic* 255(3), 1985, 29–38

Midwest Research Institute, "Resource Recovery: The State of Technology," Report for the Council on Environmental Quality, 1979

OECD, *Economic Instruments of Solid Waste Management,* Paris, 1981

Owen, E.H., "Computer Program Cuts Costs of Urban Solid Waste Collection," *Public Works,* Vol. 101, No. 1, 1970

Paessens, H., "Erfahrungen mit der Anwendung von Operations- Research-Verfahren zur Kostenoptimierung in der Abfallwirtschaft," in H.H. Hahn and H.J. Seng (eds.) *Wirtschaftlichkeit in der Abfallwirtschaft,* Inst. f. Siedlungswasserwirtschaft, Univ. Karlsruhe, Karlsruhe 1982

Plehn, St.W., "Solid Waste Management," *Encycl. of Environmental Science,* McGraw-Hill, New York, 1977

Schenkel, W. (ed.), "Abfallwirtschaft in großen Städten und Verdichtungsräumen (Waste management in big cities and agglomeration centers), *Müll und Abfall* 25, 1986, 5–91

Schlottmann, A., "New Life for Old Garbage," *Journal of Environmental Economics and Management,* Vol. 4 (1), 1977, 57–67

Wolfe, H. and R. Zinn, "Systems Analysis of Solid Waste Disposal Problems," *Public Works,* Vol. 98, 1967

Ward, J.E. and R.E. Wendell, "A Million Dollar Annual Savings from a Transfer Station Analysis in Pittsburgh," Proc. 9th Ann. Pittsburgh Conference on Modelling and Simulation 9, 1978, 1305–1308.

Yakowitz, H. and R.E. Chapman, "Evaluating the Risks of Solid Waste Management Programs: A Suggested Approach," Resources and Conservation 11, 1984, 77–94.

D. DATA BASES AND INSTITUTIONAL SOURCES

Office of Solid Waste, Environmental Protection Agency, *Solid Waste Facts,* EPA, Washington, D.C., 1978

Abfallbeseitigungsplan, Teilplan: Hausmüll und hausmüllähnliche Abfälle, Bekanntmachung des Bayer. Staatsministeriums für Landesentwicklung und Umweltfragen vom 22.5.1978 (LUMMB/S.71)

Fortschreibung des Abfallbeseitigungsplans, Teilplan: Hausmüll und hausmüllähnliche Abfälle, Bekanntmachung des Bayer. Staatsministeriums für Landesentwicklung und Umweltfragen vom 21.2.1980 (LUMBI pp. 108)

Stand der Abfallbeseitigung in Bayern 1977, Hausmüll und hausmüllähnliche Abfälle, Schriftenreihe Abfallwirtschaft, Heft 5, Bayer. Landesamt für Umweltschultz, Oldenburg Verly, Wien, 1978

Vollzug des Abfallbeseitigungsgesetzes und des Bayerischen Abfallgesetzes; Hinweise zur abfallrechtlichen Überwachung der Einrichtung und des Betriebes von Deponien für Hausmüll und hausmüllähnliche Abfälle; Bekanntmachung des Bayer. Staatsministeriums für Landesentwicklung und Umweltfragen vom 23.5.1980 (LUMBI pp. 87)

Stand der Abfallbeseitigung in Bayern 1980/81 (Hausmüll und hausmüllähnliche Abfälle), Bayer. Landesamt für Umweltschutz, 1983

Projektgutachten zur Neuordnung der Abfallbeseitigung in der Region München, Teil I, II und III, Messerschmitt-Bölkow-Blohm, Dorsch Consult, Keller und Knappich, München, 1973

Office of Technology Assessment, U.S. Congress, Facing America's Trash: What Next to Municipal Solid Waste?, Washington, D.C., June 1989.

E. ORGANIZATIONAL STUDIES

Bennett, J.T. and M.H. Johnson, "Public versus private provision of collective goods and services: garbage collection revisited," *Public Choice,* Vol. 34, 1979, 55–63

Borcherding, Th.E., W.W. Pommerehne, and F. Schneider, "Comparing the efficiency of private and public production," *Zeitschrift f. Nationalök.* (Jour. of Economics). Suppl. 2, 1982, 127–156

Hirsch, W.Z., "Cost functions of an urban government service: refuse collection," *Review of Economics and Statistics,* Vol. 47, 1965, 87–92

Kemper, P. and J.M. Quigley, *The Economics of Refuse Collection,* Ballinger Publ. Comp., Cambridge (Mass.), 1976

Kitchen, H.M., "A statistical estimation of an operation cost function for municipal refuse collection," *Public Finance Quarterly,* Vol. 4, 1976, 56–76

Pier, W.J., R.B. Vernon, and J.H. Wicks, "An empirical comparison of government and private production efficiency," *National Tax Jour.,* Vol. 27, 1974, 653–656

Pommerehne, W.W. and B.S. Frey, "Public vs. private production efficiency in Switzerland: a theoretical and empirical comparison, in V. Ostrom and F.P. Bish (eds.), *Comparing Urban Service Delivery Systems, Urban Affairs Annual Review,* 12, 1977, 221–241

Savas, E.S., "Evaluating the organization of service delivery: solid waste collection and disposal," *Waste Age,* Vol. 39, 1975, 4–14

Savas, E.S., "Policy analysis for local government: public vs. private refuse collection," *Policy Analysis,* Vol. 3 (1), 1977, 49–74

Savas, E.S., "An empirical study of competition in municipal service delivery," *Public Administration Review,* Vol. 37, 1977, 717–724

Savas, E.S., "On equity in providing public services," *Management Science,* Vol. 24, 1978, 800–808

Savas, E.S., "New directions for urban analysis," *Interfaces,* Vol 6, 1975, 1–9

Spann, R., "Public versus private provision of government service," Chap. 4 in T. Borcherding (ed.), *Budgets and Bureaucrats,* Duke Univ. Press, Durham, N.C., 71–89, 1977

Stevens, B.J., "Service Arrangement and the Cost of Residential Refuse Collection," Chap. 8, in Savas, E.S. (ed.) *The Organization and Efficiency of Solid Waste Collections,* Lexingston Books, Lexingston (Mass.), 1977, 121–138

Stevens, B.J., "Scale, Market Structure and the Cost of Refuse Collection," *Review of Economics and Statistics,* Vol. 66, 1978, 438–448

F. ECONOMICS OF SOLID WASTE MANAGEMENT

Nerlove, M., "Recent Empirical Studies of the CES and Related Production Functions," *Theoretical and Empirical Analysis of Production,* ed. Murray Brown (New York, 1967), p. 82

Nerlove, M., "Estimation and Identification of the Production Function" (Chicago, 1965), p. 106

Marks, David H. and Liebman, J.C., "Mathematical Analysis of Solid Waste Collection," H.E.W., U.S.G.P.O. (Washington, D.C., 1972)

Shepard, R.W., *Theory of Cost and Production Functions,* Princeton University Press, Princeton, NJ, 1970

G. MATHEMATICAL AND REGIONAL MODELS

Bellmore, M., J.C. Liebman and D.H. Marks, "An Extension of the (Szwave) Truck Assignment Problem," *Naval Research Logistic Quarterly* 19, 1972, 91–99

Brill, E.D., "The use of optimization models in public sector planning," *Management Science,* Vol. 25, 1979, 423–432

✓ Clark, R.M. and J.I. Gillean, "Analysis of solid waste management operations in Cleveland, Ohio: a case study," *Interfaces,* Vol. 6, 1975, 32–42

Clark, R.M., *Analysis of Urban Solid Wastes Services: A System Approach,* Ann Arbor Science, Ann Arbor, MI, 1978

Clayton, K.C., J.M. Hure, *Solid Wastes Management: The Regional Approach,* Ballinger Publ. Comp., Cambridge, MA, 1973

Geoffrion, A.M., "The Purpose of Mathematical Programming is Insight, not Numbers," Western Management Science Institute, Working Paper No. 249, June 1976

Geoffrion, A.M., "A Priori Error Bounds For Procurement Commodity Aggregation in Logistics Planning Models," Western Management Science Inst., Working Paper Nor. 251, June 1976

Greenberg, M.R., et al., *Solid Waste Planning in Metropolitan Regions,* The Center for Urban Policy Research, Rutgers Univ., New Brunswick, N.J. 08903

Handler, G.Y. and P.B. Mirchandani, *Location on Networks: Theory and Algorithms,* The MIT Press: Cambridge, MA 1979

Hasit, Y. and D.B. Warner, "Regional Solid Waste Planning with WRAP," *Jour. of Environ. Engineering,* Proc. of the ASCE 107, 1981, 511–525

Jenkins, L., "Parametric Mixed Integer Programming: An Application to Solid Waste Management," *Management Science* 28, 1982, 1270–1284

Lauria, D.T., "Regional Sewage Planning by Mixed Integer Programming," Dept. of Environmental Science and Engin., School of Public Health, Univ. of North Carolina, Chapel Hill, 1975.

Liebman, J.C., "Some simple-minded observations on the role of optimization in public systems decision-making," *Interfaces,* Vol. 7, 1976, 102–108

Ross, G. Terry, "A Branch and Bound Algorithm for the Generalized Assignment Problem," *Mathematical Programming* 8, 1975, 91–103

AUTHOR INDEX

113

SUBJECT INDEX